ISBN 978-0-428-23738-7
PIBN 10315552

This book is a reproduction of an important historical work. Forgotten Books uses
state-of-the-art technology to digitally reconstruct the work, preserving the original format
whilst repairing imperfections present in the aged copy. In rare cases, an imperfection in
the original, such as a blemish or missing page, may be replicated in our edition. We do,
however, repair the vast majority of imperfections successfully; any imperfections that
remain are intentionally left to preserve the state of such historical works.

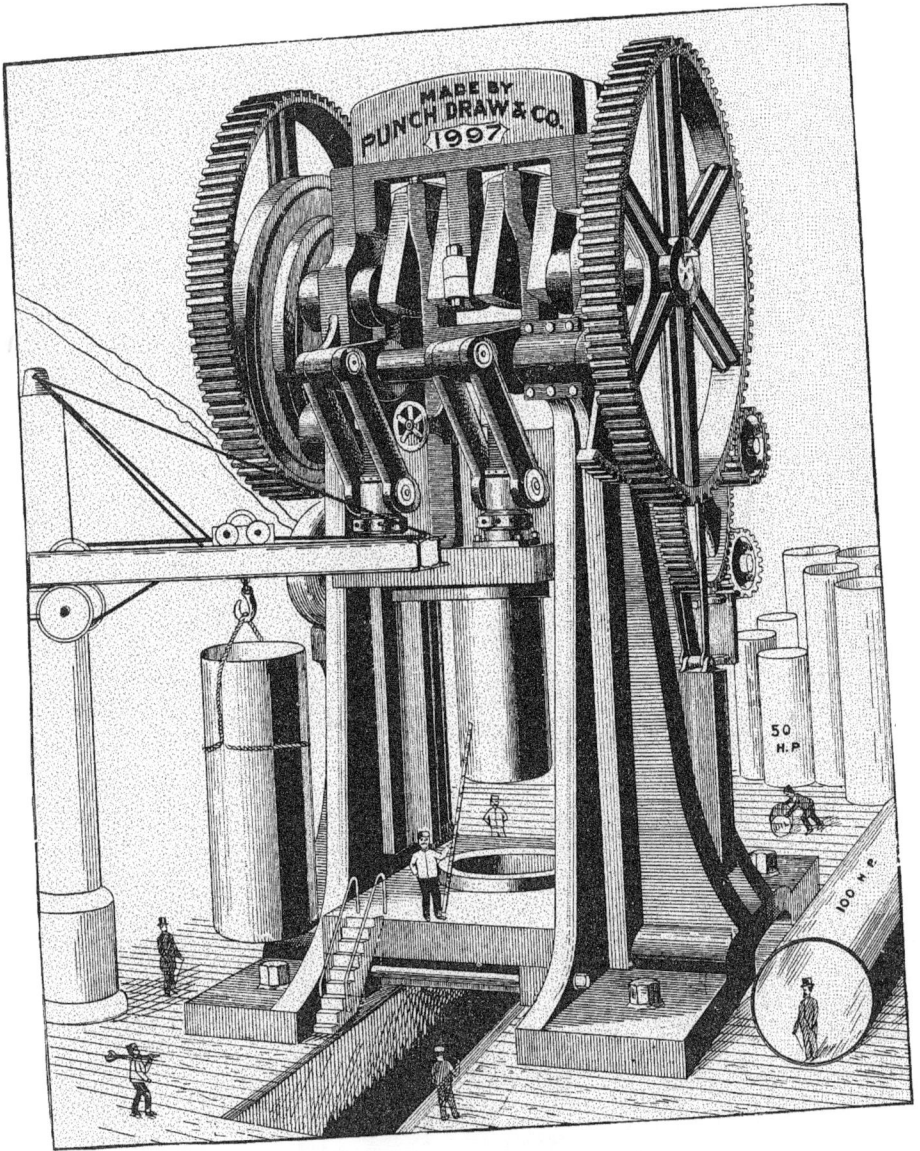

DRAWING UP BOILER SHELLS.

A PRESS OF THE FUTURE.

DIES

AND

DIE MAKING.

BY

J. L. LUCAS.

FIRST EDITION.

1897.

8-9474

PRINTED BY JOURNAL OF COMMERCE CO., PROVIDENCE, R I

TO

ELISHA T. JENKS,

WHOSE EARNEST EFFORTS TO INSTIL A FEW MECHANICAL IDEAS

INTO THE BRAIN OF THE AUTHOR WERE MET WITH

BUT INDIFFERENT SUCCESS, THIS BOOK

IS RESPECTFULLY DEDICATED.

PREFACE.

In putting this work before the public, the author is well aware that it will meet with criticism from those who are better informed on the subject than he; but a careful search in the bookstores and through the catalogues of the largest publishing firms failing to show a work of this kind, has led him to take up the task; and, although better fitted by experience for the file than the pen, has done the best that he could. It is essentially a shop book, written by a shopman, and intended for shop use.

Many of the dies shown were made by him personally, others were made from sketches, and under his personal supervision, while others were made and used in various pressrooms of which he had an intimate knowledge; and the rest, especially those on bicycle tools, are from the columns of the *American Machinist* and other papers devoted to mechanics.

Certain parts of these articles have been published in the columns of the *American Machinist;* also a few pages can be found in the catalogue of the Mossberg & Granville Manuf'g Co., which were contributed by the writer.

Trusting this work may prove a help to the great army of die-makers now pushing their way to the front, is the wish of

THE AUTHOR.

PROVIDENCE, R. I.,
    September 30, 1897.

THE ORIGIN OF DIES.

The origin of dies is, without doubt, very obscure. Perhaps the frontispiece in Mr. Oberlin Smith's excellent work on "The Press Working of Metals" will give as good a likeness of the original die-maker as we shall ever find. A search in some of the best libraries and museums has failed to reveal any information that is at all ancient, and that can be relied upon. The samples of old coins that have been handed down to us, show that the art of die-sinking was known to the ancient Greeks at least eight hundred years before the Christian era, but they do not show that the use of punches and dies was equally well known.

The first record of punches and dies used in a machine or press (having guides or ways to ensure the punch entering the die correctly) is in the fifteenth century, when they were used in the manufacture of hinges, by a German locksmith. Later on, in the sixteenth century, we find them in use both in Germany and France; and as early as the year 1796 a patent was granted to one "De Vere," of France, for dies for punching and drawing of sheet metal.

In 1827, M. Gagalott took out a patent for press and tools for drawing up cartridge shells from sheet copper. These were nearly the same shape as those in use at the present time, and the dies were probably of the same design.

In the article on "Drawing Presses," on page 22, will be found a history of the first drawing press made in this country, used for drawing up kitchen utensils,—the facts of which can be vouched for, and were related to the writer by one of the workmen employed, Mr. J. L. Kirby, now superintendent of the Dover Stamping Co. The facts relating to the use of drop presses for "striking up" sheet metal were also obtained from the same source.

The author would be pleased to learn any facts relating to the use of dies prior to that time, and would like to hear from any of his readers on this subject.

PROPER SELECTION OF STEEL FOR DIES.

Steel intended for punches and dies should in all cases when possible be ordered directly from the mill, giving all the information possible as to the use to which it is to be put, and the class of work it is intended to punch. The steel should be annealed at the mill, as the extra cost is more than offset by the ease which it is worked, and the certainty with which it can be hardened.

Much of the trouble now experienced with die steel could be avoided by this method, except that caused by improper heating by the smith in hardening. Steel for dies should be free from seams and flaws and of even color, not to be high in carbon (from .007 to .009 per cent. is a good proportion) and capable of being hardened without shrinking or changing its shape to any great extent, except in the case of drawing dies, where this fault is a desirable one, as the die can be (when worn) reduced to size by being re-hardened, thus increasing the amount of work the die will perform.

When you have once found a grade of steel that you know to be suited to your work, "stick to it" (and don't be induced by a smart agent with a box of cigars to give him an order even if he claims that he has a brand that will not shrink or change in the least, and would stand any degree of heat without injury). If you find that it is impossible to get any one grade that will fill the bill for all kinds of dies, mark each brand so that you or your men can tell at a glance what kind of stock it is. A good way is to paint a stripe the whole length of the bar, and a different color for each make or brand of stock. The stripe the whole length of the bar, indicates the make no matter how short the piece may be, and as shellac is usually the most abundant paint in the shop, use red, black, and white shellac, and but three grades of steel.

For drop-press work for cold dropping, use a steel that is a little high in carbon, and give the dies plenty of stock to avoid breakage. For very large dies for drop-press work and where they are to be used on thin stock, as tin or sheet iron, a die made of a steel casting and case hardened will give good results. Where a large amount of labor is necessary to produce the die as in the case of watch or clock dies, get the best steel that can be found, as it is a poor policy to try and use a poor grade of steel for intricate work.

MAKING A PLAIN PUNCH AND DIE.

Having found a brand of steel that is suitable for the work to be punched it should first be machined to the proper size and shape, which will depend on the class of work it is to do, and the style of die-shoe it is intended to be used in. A good way, and one that is adapted to a large class of work, is to bevel the edge of the bar to an angle of 10 degrees, leaving the ends as they come from the forge or saw. To get out these blanks cheaply, cut the steel up into two or three lengths, plane up and bevel the edges to the proper angle, and then saw off as wanted to the required lengths.

This style of die can be held by a key or set screws, whichever the practice may be.

The upper surface of the die should be finished on the surface grinder if there is one, or by hand with the use of the file and scraper, as the time thus spent is saved in the laying out of the die, and the work can be laid out better, and more distinctly than on the rough machined surface of the blank.

The surface of the die is now covered with what the die-maker calls "blueing," but what is really sulphate of copper, and is made by dissolving sulphate of copper or blue vitriol in water (a proportion of ten parts of water and one of crystals makes a good solution), this leaves a thin deposit of copper on the surface of the steel, leaving an excellent surface for laying out the die. The outline of the piece to be punched should now be traced on the upper surface of the die; care should be taken in laying out to avoid waste of metal, also that the grain of the stock to be punched runs in the proper direction so that it can be easily prepared for punching. A margin equal to the thickness of the metal to be punched must be allowed between the blanks, and if the stock is $\frac{1}{16}$″ or less, it is better to increase this proportion.

The die being neatly laid out from either drawing or template, all round corners should be drilled with a drill of the proper diameter to leave the corner the shape, and then reamed with a reamer of the proper taper to give the die the clearance wanted; the rest of the stock is now removed by drilling. Every die-maker has a way of his own for this part of the work; one will drill so that the holes will cut into each other and thus avoid broaching, while the next man will drill holes so that there will be $\frac{1}{32}$″ or more between

them, and then cut out the stock with a flat, square-
ended tool called a "broach." This is the best and
quickest way. The bulk of the stock being removed,
if there is a die-sinker in the shop, the die can be par-
tially finished on it, and by using a mill of the proper
taper, the necessary clearance can
be given; this will vary from ¼ to
3°. Excessive clearance is given
when it is essential that the blank
should drop from the die before the
next piece is punched.

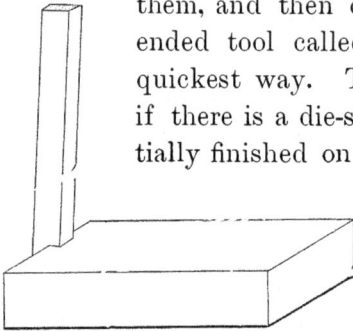

Fig. 2.

The die being worked out as far as the machine will do it, is
now finished with the file, using the die square (Fig. 2) to give it
the proper clearance, and if the shape is an intricate one that must
be closely followed, a template made of thin sheet metal and
soldered to the end of a bent wire will be found a help in making
the required shape.

If the die is for a press room using a large number of dies, it
should be marked with a shelf and rack number, so that it can be
quickly found by the die-setter; the die is now ready to harden.

The die being finished, the blank for the punch, which, if the
shape of the die demands it,
should be a forged one (but
if possible, should be cut from
the annealed bar), should be
machined on both ends, and one
end finished, then clamped
firmly on to the face of the
die, using a die clamp (Fig.
3); now scribe through the
die, and mark the outline of
the punch on the finished end
of the blank. This should be
nicely marked with a centre
punch. The stock can best be
removed by the use of a milling machine, as a base or
shoulder, of the same shape
but of a larger area, can be

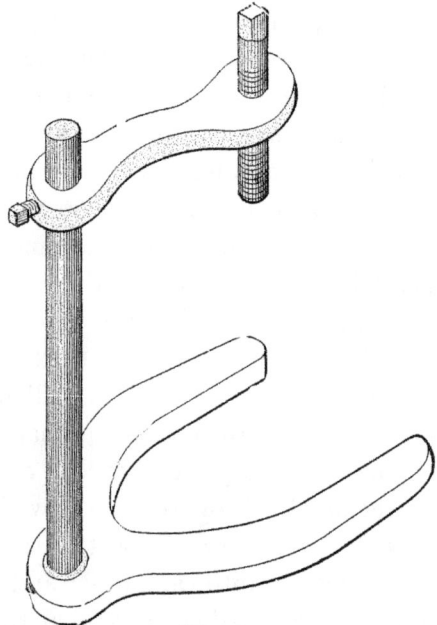

Fig. 3.

more easily left with this tool than the shaper. The object of this base is to give the punch better support, and increase the length of time it can be used.

The stock having been removed close to the lines, the punch should be placed over the die and then forced into it $\frac{1}{16}''$, thus giving a "witness" mark to file to. This is called "shearing in," and is the method usually employed in this country. The reverse of this operation, known as the "French" way, consists of finishing and hardening the punch, and then broaching out the die with it. The results obtained by this practice are not as accurate as those by the first

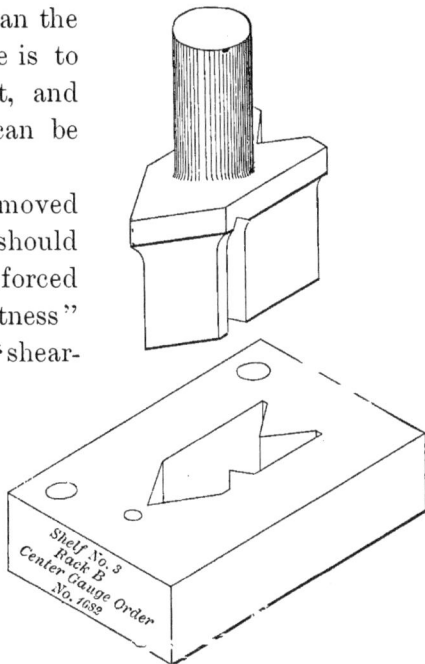

Fig. 4.

process described, as the die will change to a greater extent in hardening, than the punch, and it is not so easily finished. The punch should now be carefully filed to the lines formed by the "shearing in" process, and if the work requires it, should be hardened; but should in all cases be left softer than the die. When possible, the punch and shank should be in one piece, but if not, care should be taken to see that it has a good bearing on the punch back, and is held firmly to it.

Fig. 4 showes a plain die for blanking out a tool that all are familiar with. No stripper is shown, but the two holes at the back are intended to hold one on.

The letter "C" denotes the brand of steel that the die is made of, and is very useful in keeping a record of the results obtained from the different brands of steel.

BLANKING DIES.

A set of blanking dies consist of a male die, or punch, and the lower or female die; they are made in almost every form and size for cutting out flat blanks from steel, iron, paper, mica, etc.

Usually both punch and die are hardened and tempered; sometimes the punch is left soft, and as it gets worn, is set out and refitted by being forced or shoved into the die; sometimes it is best to reverse this operation, as in punching paper, playing cards, etc., and harden the punch and leave the die soft.

A shear is usually given to the punch or die, determined by the work it has to do; when it is intended to use the blanks, or pieces punched out, the shear should be given to the die, as less distortion is given to the metal by this method, but where the hole is the object sought,—as in rivet holes in boiler plates, etc.,—the shear should be given to the punch.

Cutting or blanking dies are usually held in a shoe or die holder, or if a large die, it is fastened to the bed of the press direct, but as a rule the fewer pieces intervening between the press and the die the better results will be obtained.

Very large blanking dies are usually made in one of two ways; either as rings set in a cast-iron holder, or by welding the rings directly on to a wrought-iron holder or die plate; the latter method is the best in making plain dies, but cannot be used in compound dies. The welding of the steel rings together, and then on to the wrought-iron plate, calls for good work on the part of the smith. Some of the finest work of this kind we have ever seen is done at the works of the E. W. Bliss Co., at Brooklyn, N. Y., large steel rings, both round and square, being welded to wrought-iron beds or die-plates, and the joints so perfect they could scarcely be detected.

In the former method the rings are first welded and then turned, hardened, and ground true in the universal grinder, then set in a groove turned in the holder as shown in the sketch (Fig. 9, page 20), and held there by being bolted from the back of the holder. The distortion of these rings, caused by long use of the die, is something that is hard to account for, the lower or die ring often being dished to such an extent as to close in the top or cutting edge of the die so much that it is necessary to grind it out and refit it to the punch, a ring of $20''$ diameter, and made of $2''$ square steel, being out over $\frac{1}{8}''$ on the bottom or lower edge; this is caused, it is thought, by the strain left in the ring in hardening, and which is released by the grinding of the die.

In making dies for hot work (as the blanking out of nuts and other thick work from red-hot metal), a die made of chilled cast-iron, with a good clearance, will give good results, as the temper is not affected by the heat of the stock punched, the punch being made of steel and fitting the die loosely (in very thick stock a difference of $\frac{1}{16}''$ or more in the diameter of the punch and die is not too much), and if a nice job is wanted, the work can be shaved, or re-punched, as it is called, by forcing it through a second die that is a trifle smaller than the first, leaving a true and smooth surface.

In the re-punching of brass and copper, the use of butter-milk as a lubricant will give a better result than any oil or soap water that we have yet found. Dies for re-punching or finishing work are not in general use to the extent that they would be if the saving that could be effected by this method were well known. Many jobs now performed on the milling machine could be re-punched, and better results obtained at less cost than by the former method.

The power required to blank out a piece of metal depends largely on the shape of the blank and the number of cutting inches in the die; a long, narrow blank requiring more power than a round blank of the same area, the shear of the dies being equal. If the work will admit of the face of the punch being slightly rounding, less pressure will be required than with a flat-ended punch.

The first thing to be considered to determine whether a punch and die should be used to produce work, is the number of pieces wanted. If it is standard work, and the demand is 100 or more per week, it is both desirable and economical to have a die made, for after the die is once made the work can be produced at a very low cost. Oftentimes when a large number of pieces are wanted, and a power feed is used, the cost will not exceed two cents per 1,000 blanks.

BENDING DIES.

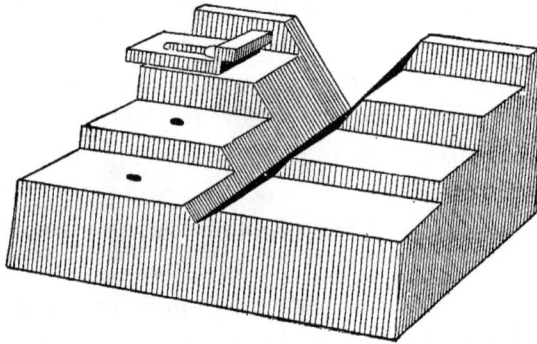

Fig. 5.

Bending sheet-iron or wire can be done to good advantage in either drop or power presses, and a few styles of dies suitable for this class of work are shown. The one shown in Fig. 5 is intended for bending any ordinary thickness of sheet metal or wire where a square bend is wanted. It is known as a Universal bending die, and will be found very useful, from the fact that it will do a very large range of bent work. The different steps are for bending an angle near the end of a strip, and for making bends of different depths.

The second die shown is for bending up a loop in a wire shown by the small sample at one side of the die. It will be found very effective in making looped wire for armature connections, switchboard work or other places where a looped wire is required. The wire is placed in the die by hand, against a stop,—not shown in the illustration,—and as the punch descends, the loop is bent down by the same, and as the end of the punch strikes the bottom of the die the spring gives way and allows the side benders,—to

COMPOUND BENDING DIE.

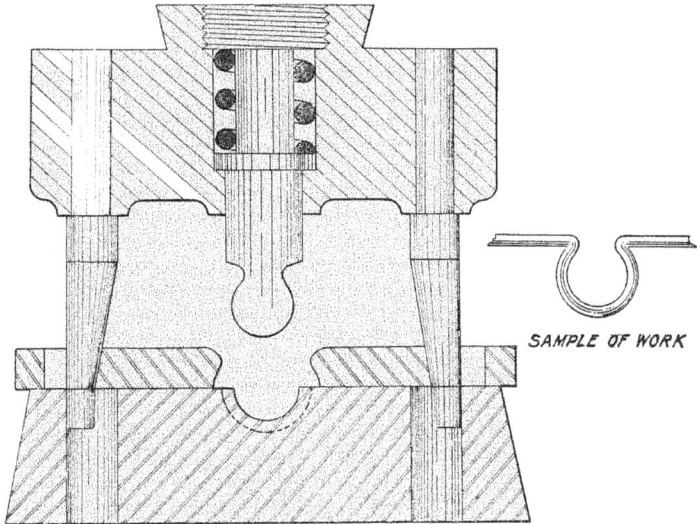

SAMPLE OF WORK

Fig. 6.

be forced in by the inclined wedges (formed on the lower end of the two pins,)—to force the wire solidly against the center punch, and as the punch rises the side bending jaws are carried back by the wire itself which is removed from the punch by hand, the end of the punch being ball-shaped to facilitate the removal of the same. The loop just formed acts as a guide for forming the next loop, the wire being moved along and the loop held against the stop (not shown) while the second loop is being formed; and then this operation is repeated indefinitely, according to the length of wire desired. This is a very successful style of die, as it does in one stroke of the press what would otherwise require a second operation.

The third die, Fig. 7, is for bending the piece on the right-hand side of the punch. This is blanked out by a previous operation placed in the die by hand, and is bent as shown by the finished piece on the left-hand side, which does not show it so clearly as it should. The two sides of the blank are bent down, and the long one on the right is curved around the punch shown,

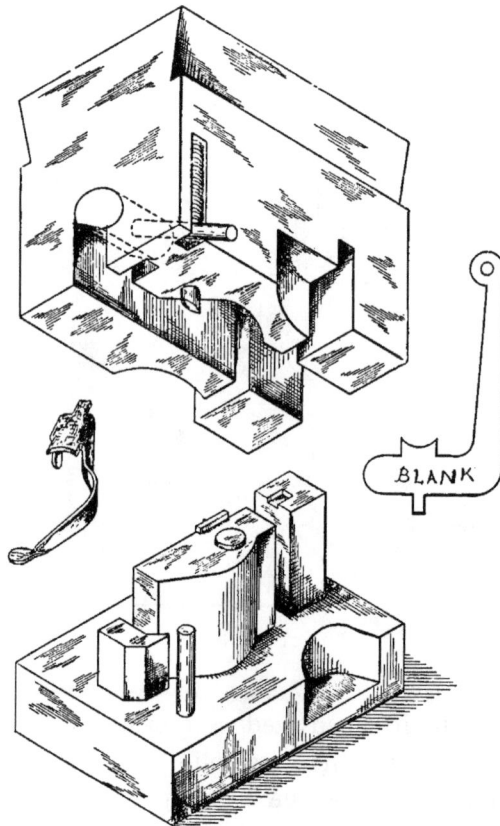

Fig. 7.

and the ring on the end is twisted around to a right angle to the rest of the arm.

The punch is made from a single block of steel which is planed up to fit the press, and then machined out as shown in the cut. The twister, which is for turning the ring at right angles to the rest of the arm, is counter-bored in and held in place by the quarter-inch pin driven into the side, and which is held up by the spiral spring seen at the upper right-hand corner of the sketch.

The die is self-contained as is used in an ordinary single-stroke press, and requires no extra attachments to enable it to do the work. This same principle can be used in many cases, and the bending performed in one operation which would otherwise require two or more strokes of the press.

COMPOUND DIES.

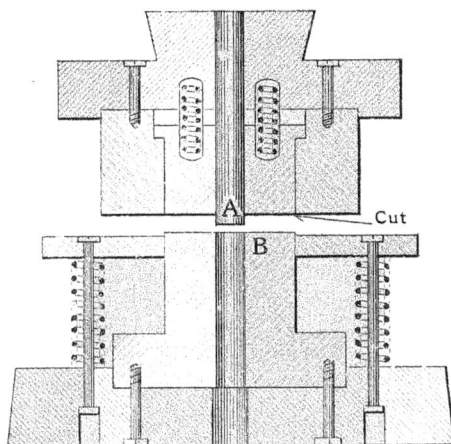

Fig. 8.

A compound die is, as the word indicates, a compound of a die and punch, or to speak more clearly, the upper half consists of a punch set inside of a die, and the lower or bottom half of a die set into a punch, the punch cutting the outside diameter of the blank, which is at the same time pierced in the centre by the upper punch, thus finishing a blank at one stroke of the press, that would otherwise require two or more operations if done on plain dies.

The work from a die of this class is far better than when done by the ordinary gang or double die, as the accuracy of the work depends on the care with which the die is made, and not on the skill of the operator. They are especially efficient when used on mica, paper and other substances that do not admit of the use of a gang die; also on work that must be practically perfect, for example—the blanks for watch and clock movements, sheet-iron disks for electric armatures, and other work where the relation of the centre or other holes to the outside diameter must be perfect.

The invention of the compound die, and the adaptation of the same to the finer grades of work, as watch and clock movements, is due to A. L. Dennison, of Waltham, Mass., and was used in connection with the sub press, also designed by him.

Though originally designed for the smallest work, the idea has been extended, and it is now used for dies of the largest size, for both light and heavy work. The die shown in Fig. 8 is a

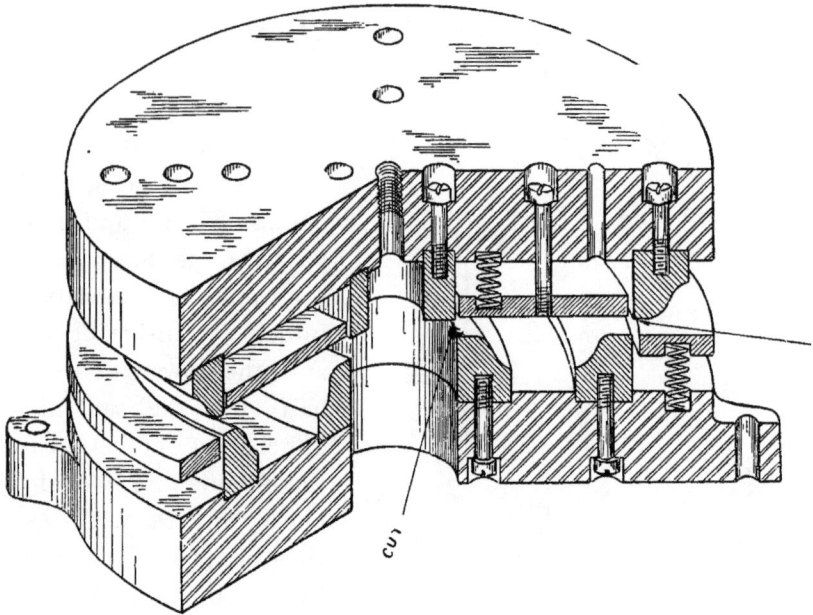

Fig. 9.

washer die, intended for cutting washers out of scrap mica, the centre punch (A) piercing the hole at the same time the outside is being cut by the die, and as a perfect washer is punched out of scrap mica at each stroke of the press, its advantages can be easily seen.

The above illustration shows a method of making plain compound dies for electrical work that will be found very serviceable —dies that can be easily made and kept in order. The upper and lower cutting rings are forged up and welded, finished in a lathe, and ground to size after being hardened.

The centre punches are made from solid stock in the same manner. Both inside and outside strippers are shown, for removing the scrap, and are made from machine steel; the whole die is seated into a cast-iron holder. The illustration shows the construction of the die so plainly that further description is unnecessary.

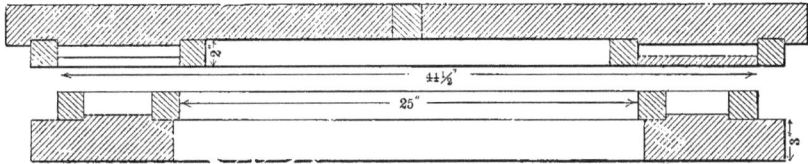

Fig. 10.

Fig. 10 is a sketch of the largest compound round die of which we have any knowledge, but should any of the readers of this book know to the contrary, we should be pleased to hear from them.

There are many dies containing more cutting surface, but in actual diameter, we think this exceeds them; it is used for blanking sheet-iron disks for the armatures of large electric railway generators for street car service:

Outside diameter . . . . 45 in.
Inside   "   . . . . 25 in.
Weight of Punch . . . . . . 1,200 lbs.
Weight of die . . . . 1,100 lbs.
Cutting surface . . . . 290 in.

In work where the stock is too heavy to admit of the use of springs, the press is fitted with a "knock-out" for the lower and upper dies, thus doing away with the springs. The pressure required to strip a complicated piece of work from the dies is very great; and a test made with a die of 16" diameter, and containing 274" of cutting edge, the force required was between 3,500 and 4,000 lbs.; thickness of stock, .025" sheet iron.

DRAWING PRESSES.

The use of presses for shaping or "drawing up" sheet metal is of comparatively recent origin, and although it was not original with the French mechanics, it was improved, perfected and first put to practical use by them, for while evidence can be found of their being in use in Germany in the sixteenth century, there are no details that are worth publishing.

In 1796 a patent was granted to one DeVere, for a press and dies for working sheet metal, and in 1827 M. Gazalott took out a patent for a cartridge shell drawn up from sheet copper. These were nearly the same shape as those in use at the present time, and probably the drawing tools were of the same design as those now made.

In the early '60's, two French workmen came to this country bringing with them a wooden model of a drawing press. This had been made from drawings taken from a press upon which they had been employed in France. This wooden press was secreted in a barn near the city of Wilmington, Del., and was finally shown to Henry Marchand, who was a brother-in-law of one of the men. Marchand, who was a mechanic of fine ability, saw at once the value of the invention, and, as he afterward expressed it, "would rather have seen the press than had five hundred dollars in gold."

Mr. Marchand at once started to raise funds to build and operate the press, and after many delays a company was formed, known as the Higgins & Marchand Co. Work was commenced on the press at Philadelphia, but Mr. Marchand soon discovered that they were secretly copying each part of the press as fast as it was made, and the parts of the half finished press were taken to Wilmington, and finished at what was afterward known as the Higgins & Marchand shop. It was set up in one corner of the shop, which was boarded in, and only three men were allowed in the room, to one of whom, Mr. J. L. Kirby, now Superintendent of the Dover Stamping Co., Cambridge, Mass., the author is indebted for this information. The press was a single action cam press, the cam being used to force the drawing punch through the die, the blank holder being held down to the work with a 3,000 lb.

weight which was worked by a long 16′ lever, extending through the wall. The first piece of work drawn up was a wash basin, made from a 14″ blank, and was probably the first piece of drawn work ever made in America.

The second press was a great improvement over the first, the long lever and weight being discarded, and side cams for the blank holder substituted in its place. This press is still in use in the shops of the Dover Stamping Co. The blank holder has been removed, and the top hamper taken off; the press is now used for a few special jobs. No punching was done on the press, as it was used for drawing only. It had been in constant operation for nearly forty years, when it was displaced by one of E. W. Bliss's latest presses, about a year ago. This is an authentic, if not an exhaustive, account of the origin of the drawing press in this country.

There is one thing about drawing up a long shell out of a round blank that is a puzzle. That is, if you draw up a long tube 1″ in diameter, out of, say, a 12″ round blank, where does the edge of the blank go? It is 36″ around that piece of metal when you start to work on it, and only 3″ when you get through. Where has the 33″ of edge gone? "It is," as one of the boys said one day, "all crumpled up and drawn out into the shell."

DRAWING DIES.

Drawing dies may be divided into three kinds, plain drawing, or "push through" dies, solid bottoms or "knock-out," and combination dies. These last named may be subdivided into single, double or triple action, or the number of operations may be extended as far as the ingenuity of the die-maker will carry him. In the private press room of the Plume & Atwood shops at Waterbury, Conn., seven distinct operations are performed on one piece of work before it leaves the press, the stock being fed in automatically, and the work carried from one set of dies to the next in the same manner. Such presses are only possible when the work is to be done in large quantities, and the cost of the press and dies is very great, three of the presses spoken of costing nearly $30,000.00.

PLAIN DRAWING DIES.

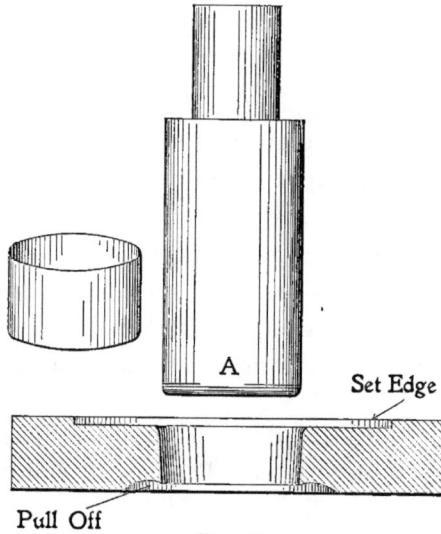

Fig. 11

In sketch Fig. 11, we show a plain "push through" drawing punch and die, the blank being punched out to fit the "set edge" and then "drawn up," or rather, it is pushed through the die by the punch, and as the punch rises, the work is stripped from the punch by the "pull off," which is made very sharp for that purpose. A "draw" of about one-quarter to one-half of a degree is given the die, making it that much larger on the upper side or face, and the upper edge is rounded over and left very smooth, and as hard as fire and water will make it. "Don't draw a drawing die." The lower end of the punch is rounded and left in the same shape; often the die will work better if the finish is changed from a circular to a lateral polish. The diameter of the punch is equal to that of the die, minus double the thickness of the stock to be drawn. A die of this kind can only be used on shallow work, or in redrawing or reducing the diameter of the work that has been previously drawn up; if used on deep drawing it will pucker or crimp around the edge. To avoid this, we must have a blank holder to hold the stock firmly while it is being drawn.

SINGLE ACTION COMBINATION DIES.

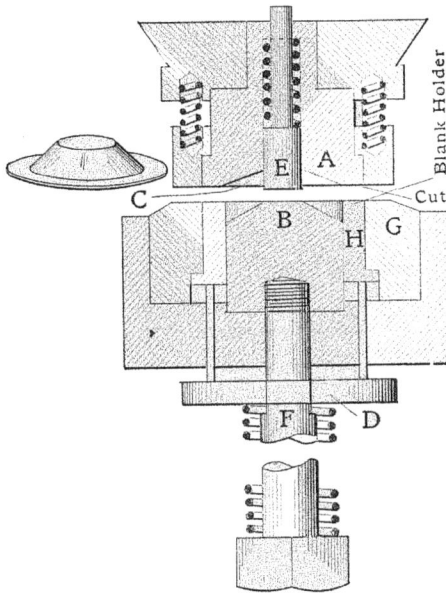

Fig. 12.

Fig. 12 shows a single-action cutting and drawing die, better known as a single-action combination die. A combination die is, as the name indicates, a combination of a drawing die and cutting die in one; it punches the blank, and at the same stroke of the press, draws it up into a cup or shell. The die shown is intended to be used in an ordinary single-stroke power press, and will draw up work not over one or two inches deep.

SOLID BOTTOM DIES.

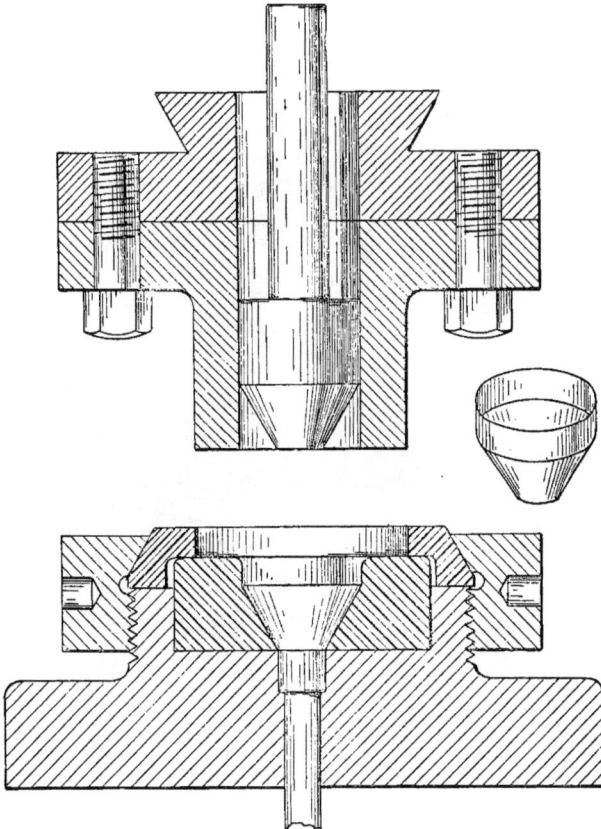

Fig. 13.

The die shown in Fig. 13 is the same general design as the previous one, except that it is a "push back" or "solid bottom" die, and is used for taper and flared work that could not be done in a "push through" die, such as thimbles, taper ferules and lamp burner parts. The action is the same as the "push through" style, except that the work is pushed back by a "follow plate" worked by a "knock-out" attached to the press, and so arranged, that, as the drawing punch rises from the work, the "knock-out" strikes the plate pin and pushes the work from the die, and is removed either by a swinging arm, or by the use of an inclined press.

DOUBLE ACTION DIES.

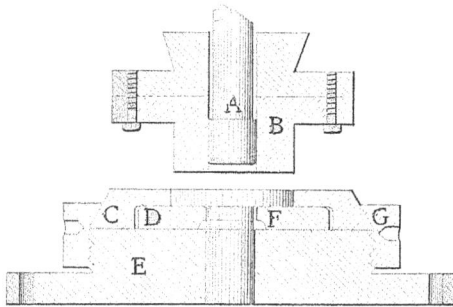

Fig. 14.

A double-action die is a modification of a single-action die, to be used in a double stroke press; it can be used on work that is too deep for single-action dies. The one shown in the sketch, Fig. 14, is known as a "push through" die. It is somewhat like the single-action die, except that the shell is cut by the punch (B), and is carried to the drawing die (D), and the lower surface of the die acting as a blank holder is held there while the drawing punch (A) forces it through the drawing die, and as the punch withdraws, the shell is removed by the lower edge of the die, which is ground very sharp for that purpose, and is known as the "pull off." The drawing die is held in place by the cutting die being clamped upon it by the ring (G).

This style of die has this advantage; the cutting and drawing dies are independent of each other, and can be changed for a longer or shorter shell, or either die can be repaired or replaced without changing the other.

TRIPLE ACTION DIES.

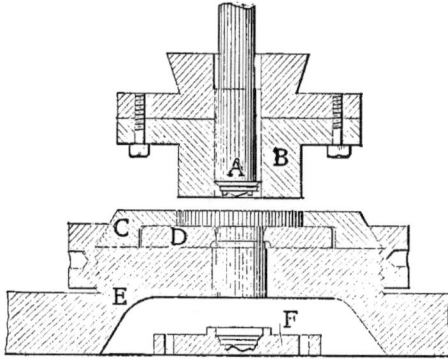

Fig. 15.

Triple-action dies are intended to punch, draw and stamp the work at the same stroke of the press. The construction is the same as the double action, except that the die block (E) is cut away to allow the stamping die (F) to be set in place, and the shell is carried down to the stamping die and "struck up" between that and the matrix formed on the end of the drawing punch; as the punch rises, the work is stripped from the punch by the "pull off," and is removed from the dies by the use of an inclined press, or by a swinging arm attached to, and operated by the press, that catches the work as it falls from the punch.

This style of die is largely used on blacking box covers, lard pail lids, or other work where a stamped or embossed cover is wanted. A die of this class should always be used in an arched press, as the strain of the stamping process is very severe on the "open back" style press, and is apt to crack the body of the press. In making a drawing die, use a steel that is high in carbon, and if it shrinks a little in hardening so much the better. There is on the market at least one grade of steel which will stand three successive hardenings, and will shrink each time so that the die as it wears out, can be "shrunk up" and then ground out to size again.

Various modifications of a drawing die are necessary in order to successfully draw up the different metals. Zinc works better when the soap suds, or whatever lubricant may be employed, is used hot as possible, as that metal works much better at a heat of about 125°.

In ·drawing up very delicate work where the nature of the metal is such that the die is inclined to clog, or as the press hand calls it "copper up," the use of butter milk will be found very effective.

In drawing very thick work the drawing die can be made bell mouthed, as the thickness of the metal will reduce the tendency to crimp or pucker. The same method of making the die is followed when it is desired to draw a shell that is very short in proportion to its diameter.

The possibilities of what can be done by this method of forming up sheet metal are almost unlimited, and the press of the future as shown by the frontispiece, is not an exaggeration of what will be performed at no distant day, as trunks, wheelbarrows, sinks and the copper boiler for hot water service are every-day productions at the present time.

The great secret of drawing up work is to have good metal, and to properly adjust the blank holder, so as to hold the metal just hard enough to prevent it from puckering. In drawing up metal, always keep one thing in mind, that is, that metal has some of the properties of water, and it will flow where it can go the easiest. A grade of steel which will harden good, a little knowledge of tool making, and a good deal of experience are what is needed to make a good drawing die.

DROP PRESS DRAWING DIES.

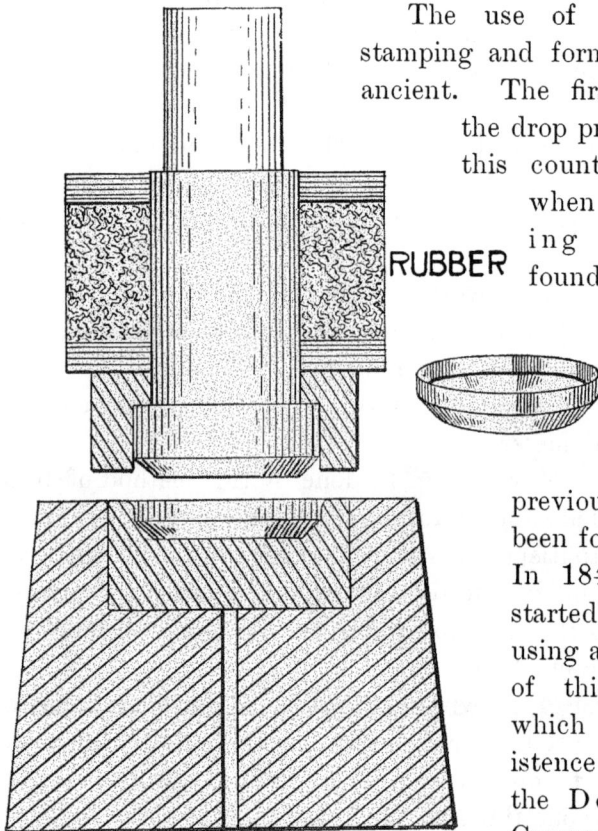

The use of drop presses for stamping and forming work is very ancient. The first information of the drop press being used in this country was in 1833, when the Dover Stamping Company was founded by Mr. Horace Whitney, at Dover, N. H. They were used for making tinware, which previous to that time had been formed up by hand. In 1847, Mr. Whitney started manufacturing, using a number of tools of this kind, some of which are still in existence at the works of the Dover Stamping Company, now at Cambridgeport, Mass.

RUBBER

Fig. 16.

These presses are used where the nature of the work demands that it shall receive a blow or force to enable it to retain the shape given it by the dies, and which cannot be readily given in a power press. They are also used for drawing up shells or cups in place of the ordinary press, as the work can be done cheaper than by the use of the power press, unless furnished with a roll or automatic feed. In the die shown, Fig. 16, the blank is held in place by the "set edge," and as the punch comes down, the blank holder strikes the work and prevents it from crimping while the punch compressing the rubber spring draws the blank down into the die, and give it the shape desired. Drop dies are also used for striking up work that has been previously drawn up in the drawing press; also the bending and forming of both sheet metal and wire. For important improvement in this class of tools see page 88.

JACK DIES.

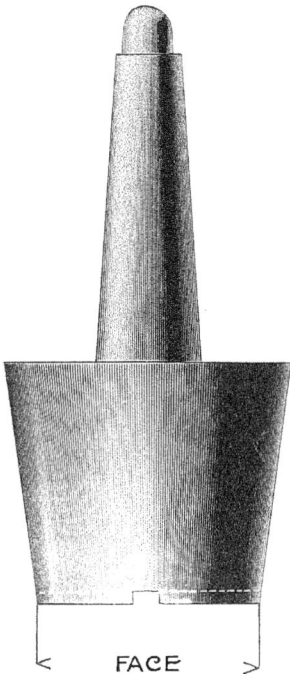

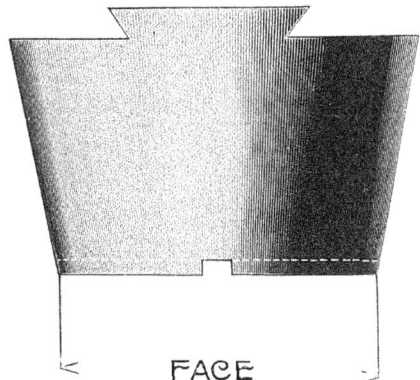

FACE

Fig. 17.

FACE

Fig. 18.

The illustrations show two methods of holding the upper die, or as it is sometimes known in shop language, the "jack" or "pick-up" die; the die being held by a taper shank which fits into the hammer of the drop, or by the dovetail which is keyed into the slot.

The former method is best adapted to hammers of 100 lbs. weight or less, and the latter to large dies or pick-ups. The under or face side of the jack is scored across with two or more slots at right angles to each other, and the face hardened. The lower die is made and finished, and the blank for the upper die or force is heated to a cherry red, and placed under the hammer which is brought down with all the force that can be applied, driving the hot metal both into the slots in the pick-up, and into the lower die, forming the punch or force, and at the same operation attaching it to the hammer.

This seems to be a crude method of holding what is often a nice piece of work, but careful search proves it to be the one universally practiced in all sections of this country.

ARMATURE DIES.

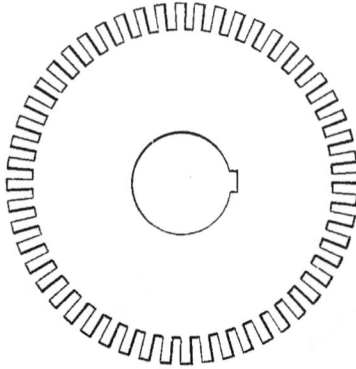

Fig. 19.

The phenomenal growth of the electrical industry in the last few years, and especially in the use of notched or slotted sheet-iron disks for the armature cores, has stimulated the invention and use of compound dies of large diameter and complicated designs.

The modern armature resembles somewhat a gear that is built up of thin sheet-iron plates or disks, usually about .010″ to .040″ thick, and from 12″ to 100″ diameter. These are, in the best of armatures, punched from sheet iron, with centre holes, keyway and the teeth, or, as they are called, "the winding slots," at one

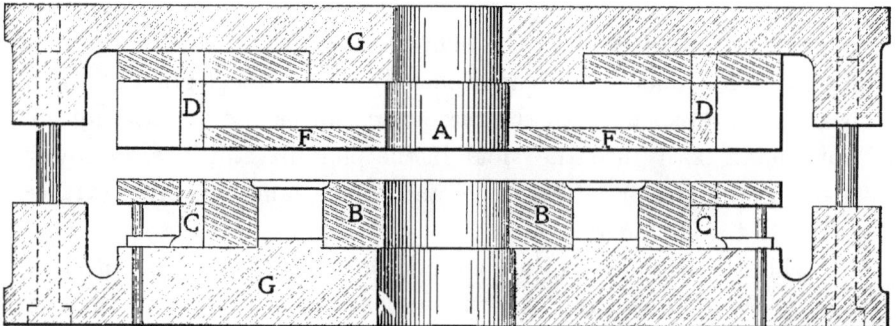

Fig. 20.

stroke of the press; the smaller ones in a complete disk, the larger sizes in as large sections as it is possible to procure the iron; and as some of the styles are very intricate, the ingenuity of the die-maker has been taxed to the utmost to produce what is wanted.

The method first adopted was to blank out the plain sheets, and then put in the centre hole and keyway with a separate die. The sheets were assembled on the shaft and turned to size on the lathe; the slots for the winding coils were then cut in a milling machine, and the core was ready to wind. It was soon found that the contact of the sheets caused by the milling process was detremental to the electrical efficiency of the armature, and the use of a dial-feed press was adopted for punching the slots before the sheets were assembled on the shaft. This gave better results, except that irregularities of the punched sheets required a good deal of filling to correct them, the effect of which was the same as that caused by the milling process, except to a less extent. This led to the development of compound dies that would punch sheets so perfectly that they could be assembled on the shaft or spider, and the armature wound without any further labor expended on it.

Fig. 20 shows one of the latest designs of these dies, adapted for use in an E. W. Bliss, No. 95 press, which comes nearer meeting the requirements called for, than any other in the market. This press is fitted with a "knock-out" capable of exerting a force of 5,000 lbs., thus obviating the use of springs in the lower stripper. The upper stripper is also worked by a stationary "knock-out" attached to the body of the press.

This die is a sectional one, is set into a steel casting, instead of cast iron, as is generally done. The rings "A," "B" and "C" are of tool steel hardened and ground to size, and the rest of the rings are left soft.

These dies are invariably sub-pressed, as the utility of the die is greatly increased by this process.

A die of this kind, well made, and run at a speed of 55 revolutions per minute, operated by two men, will produce 4,000 to 6,000 20″ sheets per day of ten hours, using 10,000 to 15,000 lbs. of iron.

The cost of these dies varies from $100.00 to $800.00 each, according to design of the armature, and they weigh from 1,000 to 2,000 lbs. each.

ARMATURE DIES.

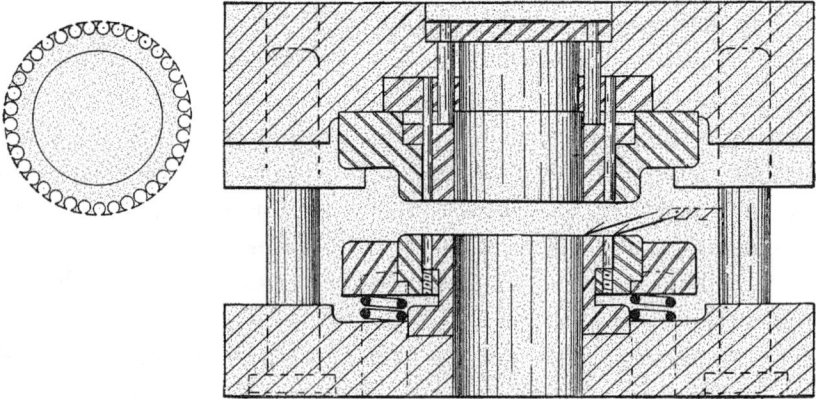

Fig. 21.

The above illustration shows a die for a small armature blank for fan motor work; also a punching from the same. The die is a compound one, sub-pressed by the use of four pins, two of which are shown in the illustration.

The difficulties of making the die were such that, at first, it was deemed impracticable; but the urgent necessity of getting out the work, which was being done by the use of a dial feed press, made it advisable to make a trial of this method of punching.

The lower die was made precisely as shown by the punched blank, of a piece of Styrian steel, which was carefully annealed, turned to the required shape, drilled for the thirty-seven holes, and then slotted on a milling machine and left soft.

The upper punch, or, as it really is, the die, was made in two pieces; the inner punch hardened and ground and forced into place in the upper casting, and the outer cutting die made in the same manner and seated in a recess machined out concentric with the centre punch.

The knock-outs or strippers were made, the lower one in two pieces, and then held together by the outer ring shown, which was forced on, holding them firmly together. ·This knock-out was fitted with thirty-seven pins of the same diameter as the holes in the disk, so arranged that as the punch was withdrawn the scrap was forced from the holes by the action of the spring shown in the illustration.

The upper knock-out, for removing the blank itself, was actuated by a knock-out attached to the body of the press, and so arranged that when the press reached the highest point of the stroke it was forced against the plate shown in the centre of the upper part of the die.

The die was a success, and was run continuously, day and night, producing several million sheets before being replaced by a new one.

NEEDLE DIES.

The illustration, Fig. 22, shows a small die and holder which is largely used in needle factories for punching the eyes of sewing-machine needles, and is intended to be used in a press adapted to that class of work. The die is a sectional one, made in three pieces, the centre one projecting up above the surface of the two sides and into the groove of the needle, thus supporting the latter while it is being punched. This centre piece is cut out with a U-shaped slot, the width of the slot being equal to the length of the eye of the needle, the two blocks forming the sides of the die, and the whole held firmly together by a set-screw not shown.

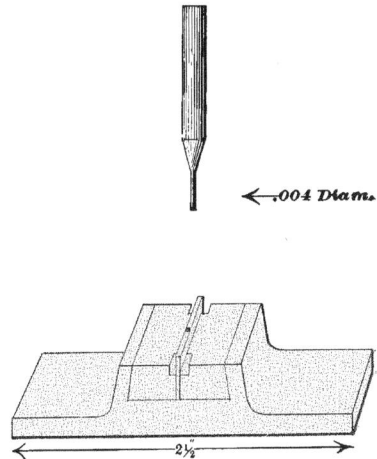

Fig. 22.

The press is so arranged that the needle is first clamped (not shown) and then punched, the speed of the punch at the moment it enters the die being very slow. The punches are made of Stubb's wire about .080 diameter, milled down to a diameter equal to the length of the punch, and are then hardened and the sides are ground in a special grinder. A punch of .004 diameter is used on the finest needles, and a punch of .008 to .010 is a common size in every-day use.

FLUID PUNCHES.

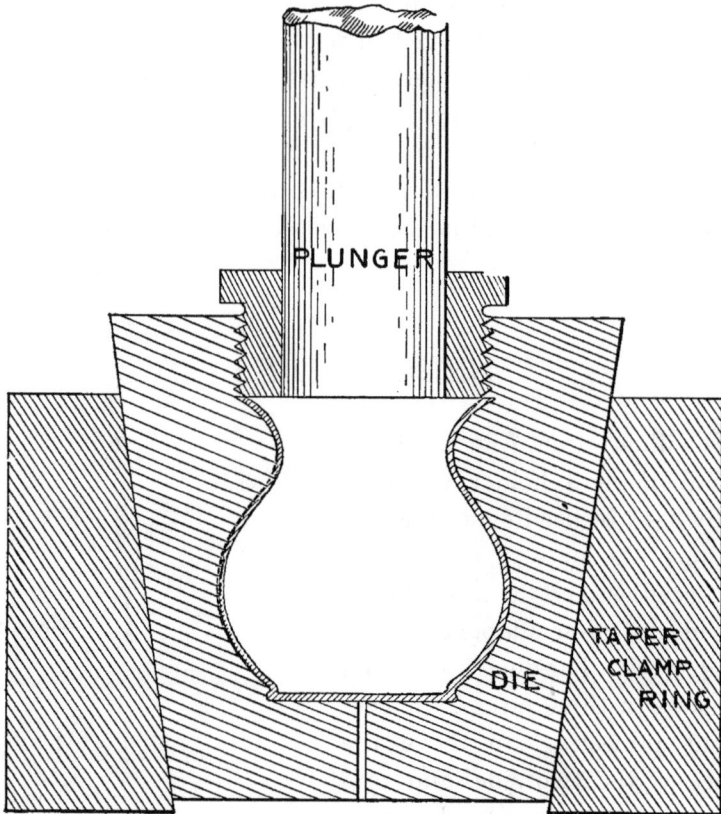

Fig. 23.

The well-known fact that water cannot be compressed to any great extent, has led to the use of fluid punches for shaping and expanding work that would otherwise be impossible to do in a press, and may be used as a medium for conveying force, as in the case of a hydraulic press, or the fluid may be applied directly to the work with or without the use of a packing ring.

The illustration shows a die designed for shaping ice pitchers, and intended for use in a heavy drop press, the die being made in two parts.

The pitcher is placed in the die, which is then closed and placed in the clamp ring, force being applied by the plunger swelling out the work to the shape of the die. A special preparation of wax is often used in place of water or oil in connection with dies of this kind. In presses used for the manufacture of hats, 800 lbs. pressure to the square inch is applied to the work. The lower die which is of cast iron or spelter, is placed on the bed of the press and then brought up to the water cylinder and locked. The water is then turned on and a bag of thin rubber is expanded, filling the entire inside of the hat and allowing the whole pressure of the water to come upon the material, remaining some ten seconds, when it is removed the press unlocked and the hat removed. When the water is applied directly to the work, it is usually done in a drop press as the pressure can be applied very quickly and but little packing is required. This method of press work is also followed in the manufacture of silverware, or where it is desired to swell the centre of the work to a diameter larger than the end of the same.

RIVETING DIES.

Where a large amount of riveting has to be done, the work can be produced better on a riveting machine arranged to spread the head of the rivet by a revolving heading tool than it can on a press, but ofttimes it is desirable to do a limited amount on a press, and the die shown will be found a very good one for the purpose. The illustration is so plain that it explains itself. It is intended for bicycle chain work, the upper and lower dies are made convex, and the rivets headed by rapid reciprocating blows of the same. A stiff short stroke press is necessary for riveting.

When a large number of holes are required to be riveted, and the work is of a nature that it can be done at one blow, the drop press will be found preferable. In a shop engaged in the manufacture of wrenches, seven to ten rivets are headed at a single blow of the drop. On page 47

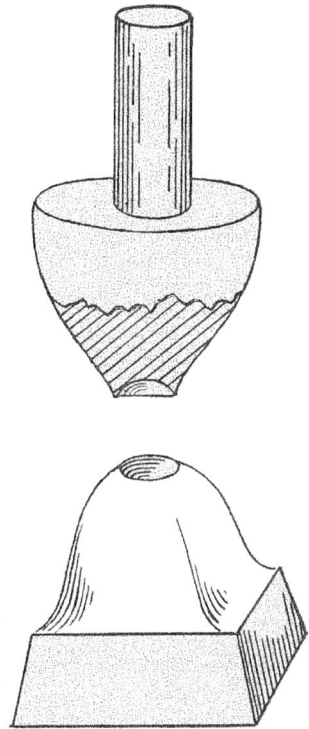

Fig. 24.

will be found an illustration of a die for imitation rivets, which, as it is not really a riveting operation, is shown separately, that it may not be confounded with the one just described.

GANG DIES.

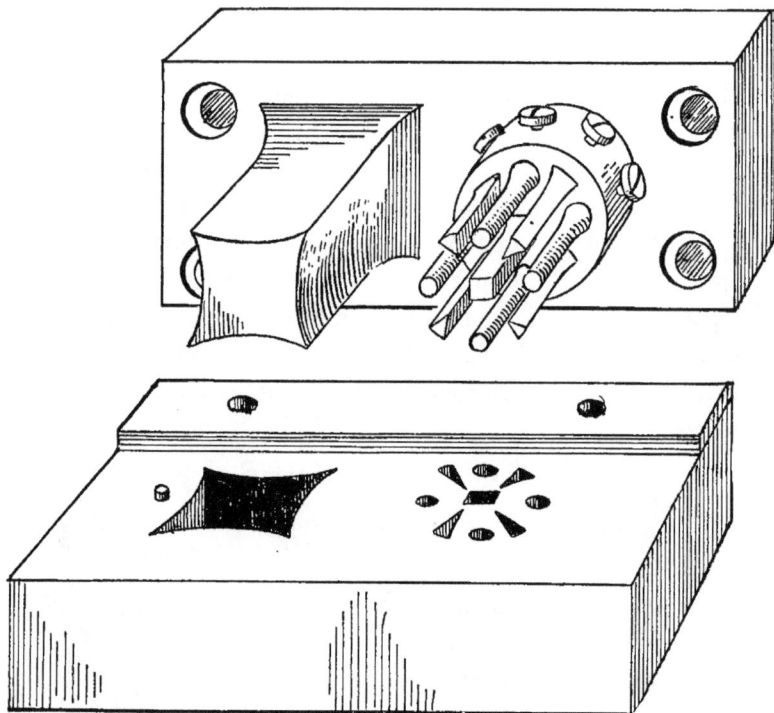

Fig. 25.

Fig. 25 shows a die intended for punching out a diamond shaped piece pierced with four round and four triangular holes, and a square hole in the centre. The die being a small one is made in one piece of steel.

Laying out of work of this kind required skill on the part of the die-maker. Proper allowance must be made for the "creeping" of the stock, also for the margin between the pieces punched. The large punch should be fitted with a centre pin projecting into the square hole which has been previously punched by the centre gang punch; the pin, however, is not shown in the cut.

When the die is a large one, and especially if the holes or work is round, it is preferable to make the dies separately, and then set them into a cast-iron holder.

Gang dies are also used very extensively when a large number of blanks of the same kind are wanted, also in punching rivet holes in tank or boiler work, and making perforations in gas or lamp burners, as high as 1,500 holes often being pierced at one stroke of the press. In work of this kind it will be found advisable to have the stripper fit the punches very tightly, so that it will act as a guide, and support the punches while they are entering the stock. A good illustration of a stripper of this kind is found on page 64.

SECTIONAL DIES.

Dies of large sizes or of a difficult shape, can usually be made to better advantage in sections, for in the former class the dies can be separately made and tempered, and set in place in the holder, thus forming a complete die, requiring less stock. Should a part be spoiled in making or hardening, it can be easily replaced; whereas, in a solid die, the whole would be spoiled.

Fig. 26 shows a method of making a small compound die in sections that was adopted because it was found impossible to produce work that would meet the requirements of the case, with the best solid die that could be made. The pieces wanted were simply flat blanks of .014" thin sheet-iron 2" X 4" in size, with ⅜" centre hole. A limit of .002" was all that could be given on the dimensions. A large number of these was wanted, requiring the use of dies that must not vary over .002" from each other.

Fig. 26.

After many trials with solid dies, which failed on account of the contraction caused by hardening, also by the changing of the die as it wore out, on account of the

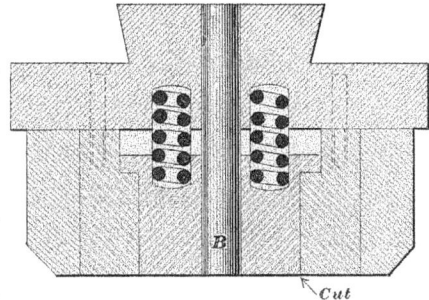

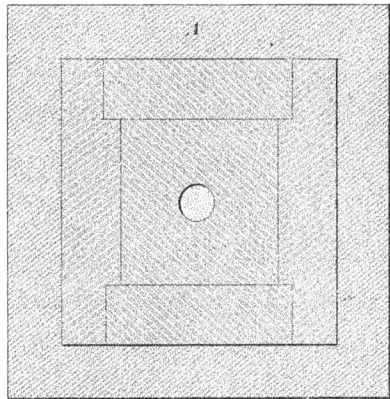

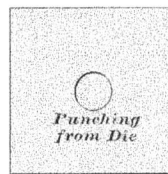

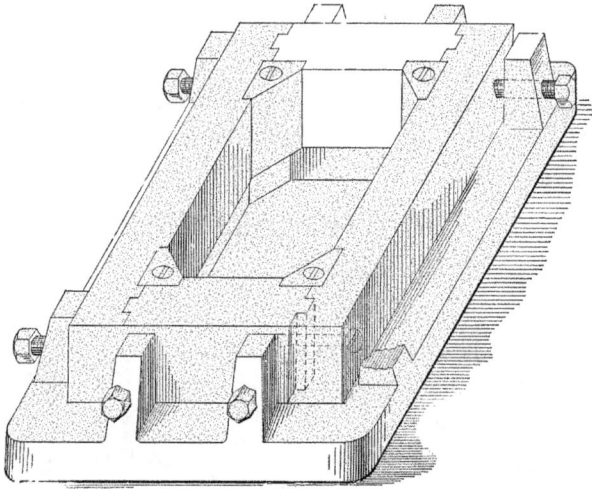

Fig. 27.

clearance given, a die was made like the sketch, the sides and ends of the die were made in separate pieces, hardened and ground to size, and held in place by the shrink ring, the centre punch "B" perforating the blank at the same time that the outside is cut. No clearance was given to the die, as the work was removed by a spring stripper, thus rendering the clearance unnecessary, and enabling the die to hold its size until worn out. This style of die gave such good satisfaction that it was used exclusively for this class of work, and the extra cost of the die was more than offset by the decreased cost of the work produced.

Fig. 27 illustrates a large punch and die that was made sectional, from the fact that it was impossible to get steel of the proper size for the die, as soon as wanted. The punch and die were both made in sections, and ground to a perfect fit after being hardened. The dotted lines show the manner in which the joint was made. The results obtained from the die, in cutting qualities and shape, were better than had it been made solid. The size of the die was about 18″ X 20″, and the stock punched was thin sheet iron. The labor cost of the work, though somewhat more than it would have been for a solid die, was more than offset by the saving in the amount of stock used. The die was set in a cast-iron holder having eight lugs which were fitted with setscrews, forcing it firmly together and making a solid die.

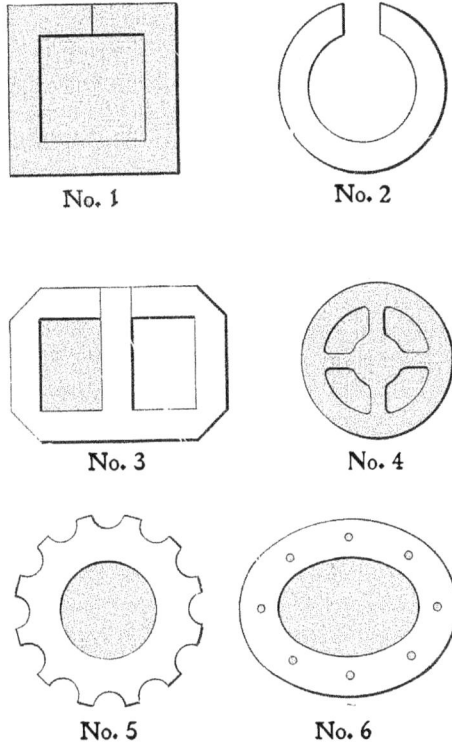

No. 1

No. 2

No. 3

No. 4

No. 5

No. 6

Fig. 28.

Dies for armature sheets are almost invariably made in sections, so arranged that each piece may be taken out and replaced without disturbing the rest of the die. Fig. 28 shows some of the shapes made in this style.

No. 1—A square sheet-iron disk with centre hole, also cut across the side.

No. 2—A washer with a section cut out.

No. 3—A transformer sheet.

No. 4—A clock wheel blank.

No. 5 and 6—Armature sheets, punched at one stroke of the press, the dies made in sections.

BURNISHING DIES.

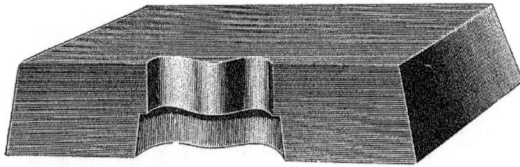

Fig. 29.

When a nicely finished piece of punched work is wanted, the piece is first blanked out a little larger than the finished size and then pushed through a die that is slightly tapered, that is, it is made larger at the top or upper side. The cutting edge of the die is a little smaller than the blank so that a shaving is taken off around the edge, and then it is forced through the die, which is made very smooth on the upper part. This leaves a finely finished surface on the edge, and is used on work that would otherwise require milling or buffing, such as bicycle chain links, and the large and small sprockets, for finishing operations. The die shown is intended for finishing bicycle links by the use of a dial feed. The centre hole is punched at the same time the link is blanked, and is centered on the burnishing die by the use of a pin that enters the hole and holds the links firmly while passing through the die.

It is important that the inside of the die is made smooth, well finished after being hardened. This style of die will, when made and properly used, save a large amount of milling and buffing, and will produce work that is uniformly accurate.

WASHER DIES.

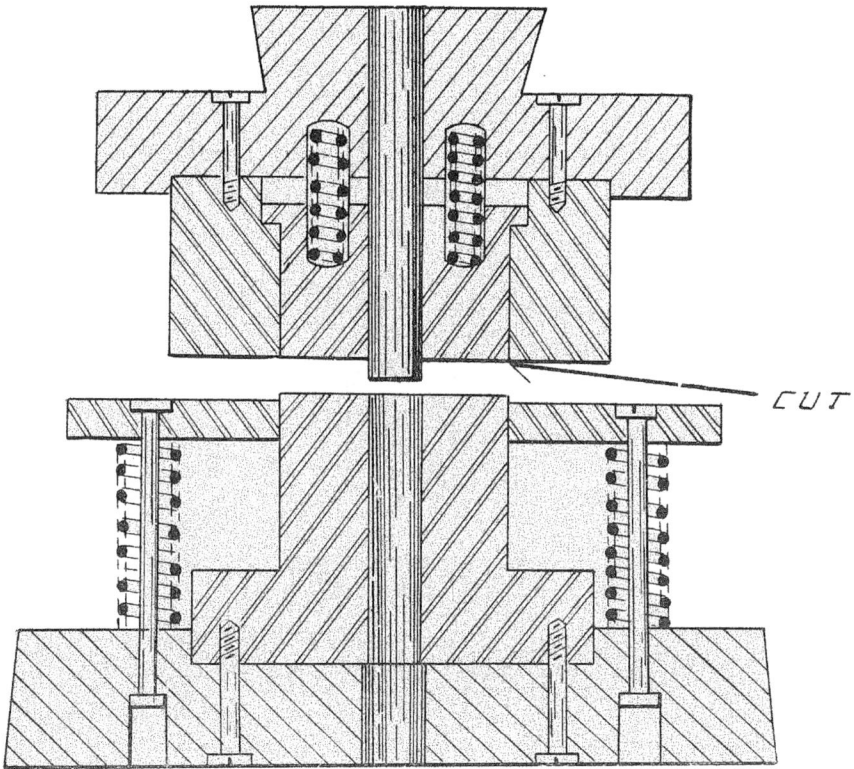

Fig. 30.

These illustrations show three distinct methods of producing washers, each of which has its advantages according to the material used.

The compound washer die in Fig. 30 is used for thin work, and where an accurate cutting die is wanted, for punching paper, mica, thin sheet-iron or brass. The illustration shows so plainly the method of making the die that little explanation is necessary. The spring stripper is used on the lower die, as it is called—which is in reality a punch—and a spring shedder or knock-out for ejecting the blank from the upper die. The advantage of a die of this kind is that the material is pressed flat and smooth by the spring stripper before being cut, thus giving a more perfect washer than can be made by any other method.

REVERSED WASHER DIE.

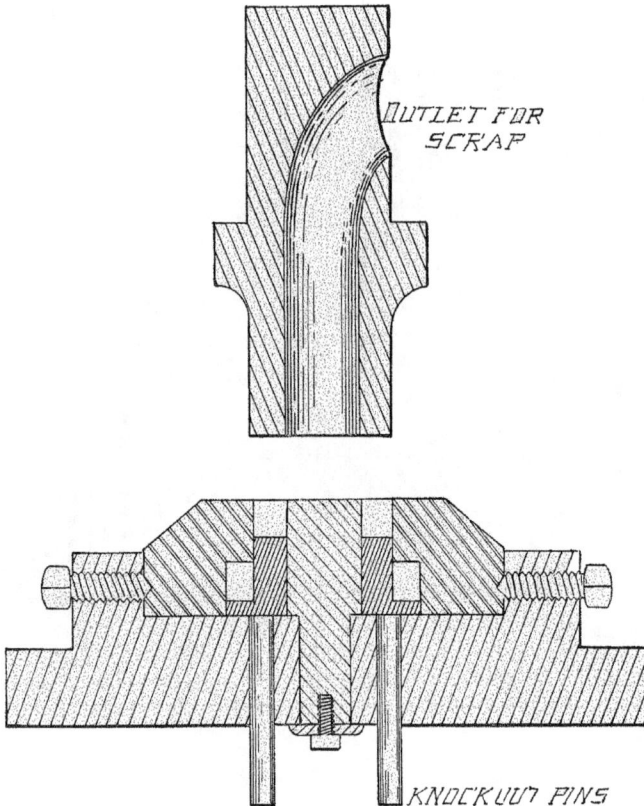

OUTLET FOR
SCRAP

KNOCKOUT PINS

Fig. 31.

The die shown in Fig. 31 is intended for thick washers, the punch being made with an outlet through the shank of the same for the scrap from the centre of the washer, the blank remaining in the die until it is ejected by the shedder which is actuated by a knock-out motion applied to the press. This die can be used for washers ¼" and ⅜" thick, and will produce perfect work even when using scrap stock, which is a difficult thing to do when using the gang washer die shown on the following page. It is especially adapted to collars and ball cups for bicycle work.

GANG WASHER DIE.

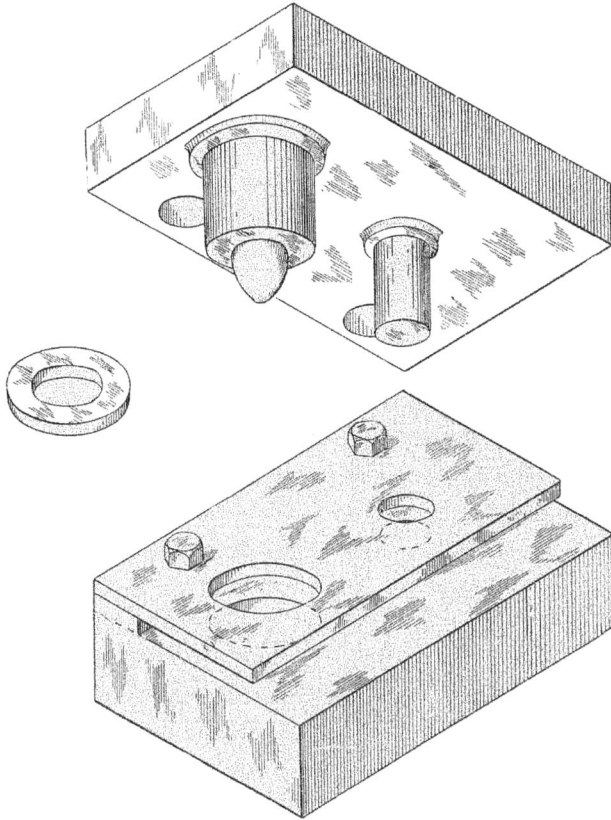

Fig. 32.

This illustration shows a gang washer die intended for punching washers of any description, and is the style ordinarily used. It is necessary in using a die of this kind that one edge of the stock be straight, in order that in feeding it may be kept firmly against the back gauge so as to ensure its being properly fed.

Another method of producing washers may be described, and consists of first cutting out the blank with a gang of punches, the press being fitted with a roll feed, and punching 25,000 washers per hour, or from one to two tons per day. These are then put through a second operation for piercing holes in the same, the press being fitted with two dies which are fed by tubes, the washer being carried to the bottom of the tube and deposited under the centre of the punch which pierces about 200 washers per minute, or about one-half ton per day.

TRIPLE DIES.

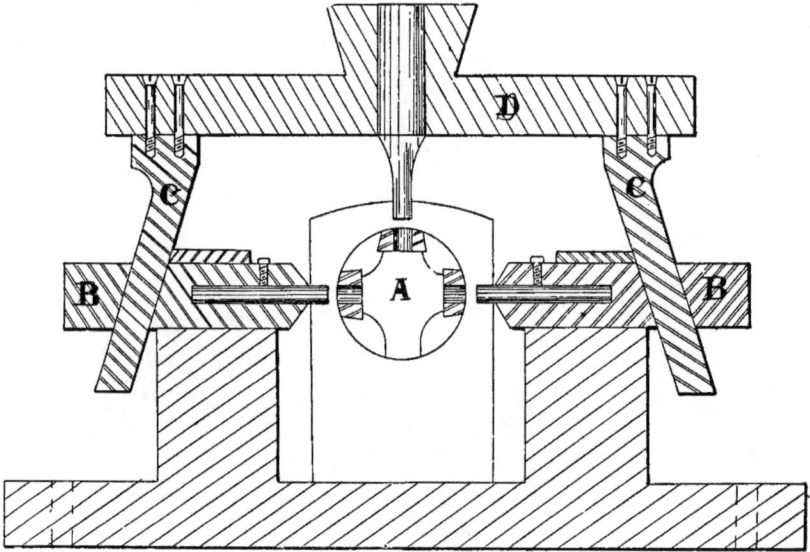

Fig. 33.

This illustration shows a die intended for slotting the ends of tubes or other shell work, and is intended to be used in any ordinary power press.

The dies are held by a hollow stud which fits the inside of the tube or shell to be punched, the scrap passing into the cavity (A) and dropping through the outlet below, left for that purpose; the punches being held in the side carriers (B), which are driven by the inclined studs (C) secured by the punch holder (D).

The die is self-contained, and can be removed from the press and set in the die rack in the usual manner. The punches are straight pieces of steel wire of the proper shape to fit the die, and are held in the same manner as shown on page 51, Fig. 39, and not by the set-screws as shown in the illustration.

The advantage of a die of this kind is the reduced cost of the work, and the better results obtained, as it is impossible to get out imperfect work if the die is properly made and used.

IMITATION RIVET DIES.

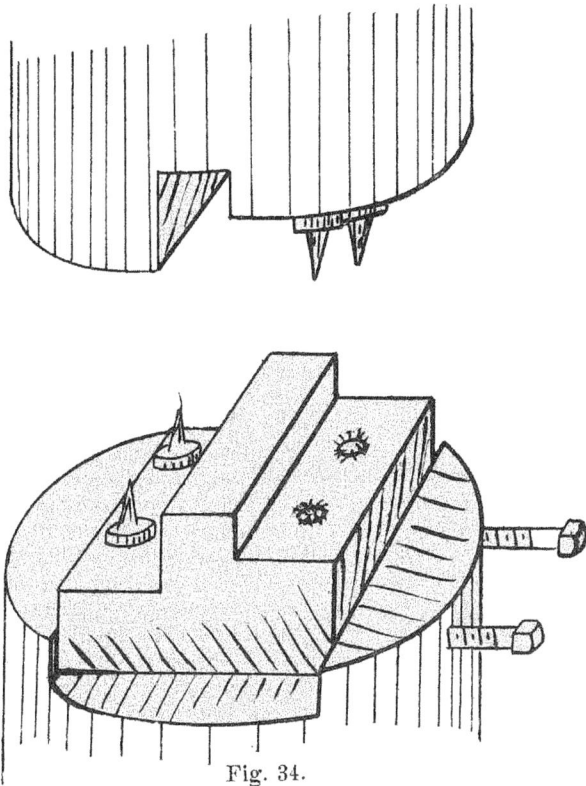

Fig. 34.

The above illustration shows a die for imitation rivets intended to be used on thin sheet-iron work, such as hoops for tubs and water pails, tinware, attaching hinges to stove ovens, or other work of a similar character.

The advantage of the die is, that no stock is removed from the parts to be joined, and no rivets are required. The punch is driven through the double thickness of stock on the first set of punches, and then finished on the second set. The only object of the second set of punches is to avoid turning the work over, as the operation could be performed equally well on a single set.

It is not really a riveting operation, being more of the nature of an eyelet than a rivet, but is described as such from the fact that it is often used in place of solid riveting, especially on thin work, such as pail hoops, band iron, and in putting handles on the cheaper grades of tinware. The operation can be rapidly performed, and applied to many classes of work not specified.

FOLLOW DIES.

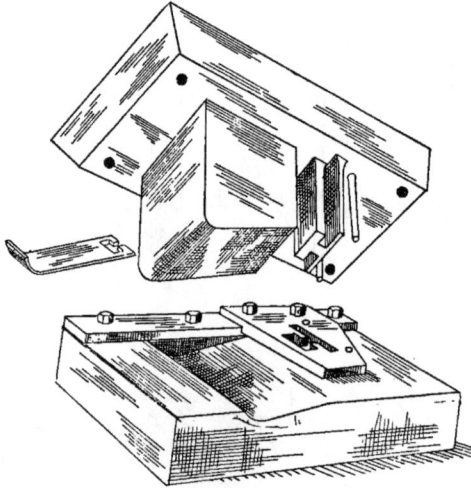

Fig. 35.

Follow dies, so called when two or more operations follow each other, that is, the piece is first stamped, pierced and then blanked out. Sometimes this order of operation is reversed and the piece is stamped and cut off at the same time, the holes having been punched at a former stroke of the press.

The illustration shows a die for making a clip, used for holding armature binding wires, and will be recognized by many of our readers. The stock is first stripped the proper width and fed into the dies by hand and the finished clip is knocked off the die in the same manner.

The same idea can be used for a large variety of work. Slots or holes can be punched in either or both ends of the blank, and any style of bend can be made that does not exceed an angle of ninety degrees, and where the depth of the bend is not more than three-eighths or one-half inch.

The difficulty of making a long bend is that the cutting punch must do the work before the bending die comes into use, and must necessarily pass down into the die whatever the depth of the bend may be. The work should be entirely cut off by the punch just at the moment it is struck by the bending die, otherwise it will be apt to be moved from its proper place, resulting in imperfect work.

ARMATURE LEAD DIE.

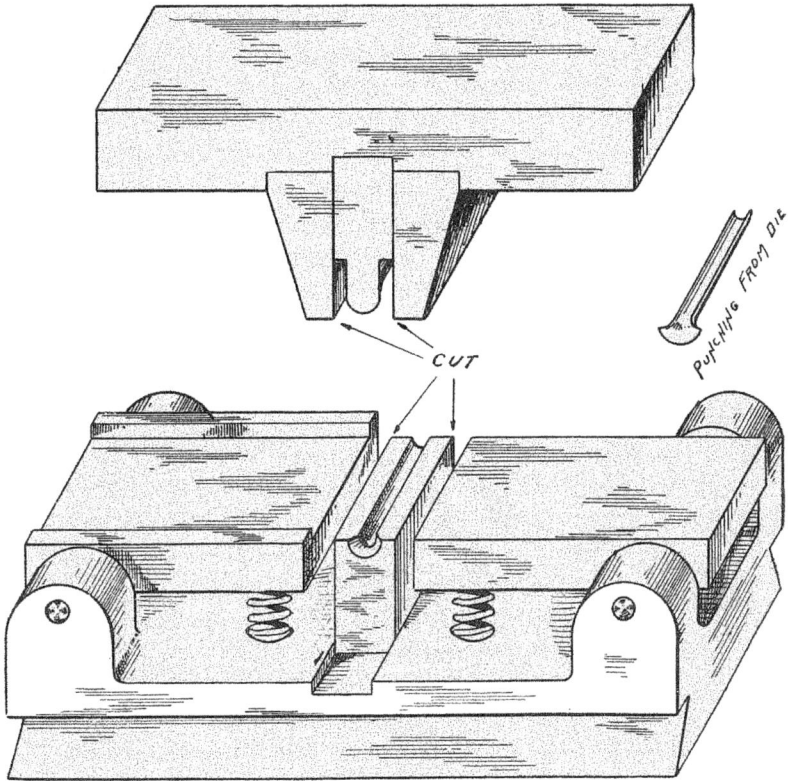

Fig. 36.

The illustration, Fig. 36, shows a die intended for cutting and forming a half-round piece, or, as it is called, a *lead*, which was formerly in great demand for armature work. The stock is fed from the left of the die, covering the small die shown in the bottom holder, and as the upper die comes down the stock is trimmed on both edges and bent to the shape shown, as the press reaches the full length of the stroke.

The piece shown at the right of the die is a gauge, and is supported on pointed screws, held up by the springs in the centre, so that as the upper knives come down it is depressed out of the way, but rises to its place again as soon as the punch returns.

This idea can be used for a large class of die work, and requires no attachments of any kind applied to the press.

TUBE BENDING DIES.

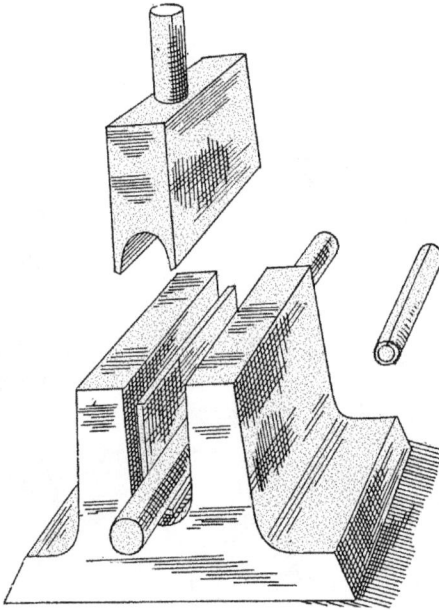

Fig. 37.

The illustrations show two methods of forming up a tube at one stroke of the press, doing the work with a single operation.

The first of these, shown in Fig. 37, is published by permission of Mr. C. H. Hardinge of Chicago, who gives the origin of the same as follows: That he obtained the same from a die-maker of his acquaintance who learned the trick from a tramping journeyman.

The second sketch is for the same purpose; the action of the dies are the same; the method of construction being changed to obviate the necessity of cutting out the deep slot shown in the first die.

These sketches are taken from the columns of the *American Machinist*, and not having seen the same in practical operation, it is impossible to state whether satisfaction equal to that obtained by the use of two sets of dies can be had by their use.

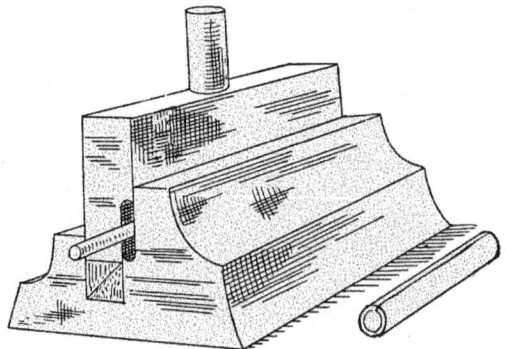

Fig. 38.

HOLDING SMALL PUNCHES.

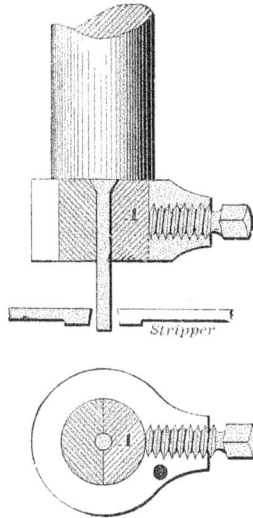

Fig. 39.

Small punches always give more or less trouble from breaking and then sticking in the punch back or holder. The method shown in Fig. 39 is designed to overcome this difficulty. The punches which are straight pieces of Stubbs' steel wire, are cut off and headed on the upper end and hardened. The chuck, which is made with a removable collar, is split in the centre, one side of which can be removed by loosening the set screw. The ease with which this can be done makes it unnecessary to send the job out of the press room. The punches can be made and kept in stock and renewed, and the punch replaced without the necessity of removing the die from the press.

Small punches are broken, not in punching the work, but in the withdrawal of the same. The cross section of the stripper in the centre of the cut shows how it should be made so as to prevent this trouble, the stripper being machined off on each side so as to have a bearing on the work to be punched, close up to the punch. When possible, the stripper should be bolted on to both sides of the die.

Out of many methods and plans devised for holding punches, this one gives the best satisfaction, on account of the ease of making and removing the punches.

SUB PRESS DIES.

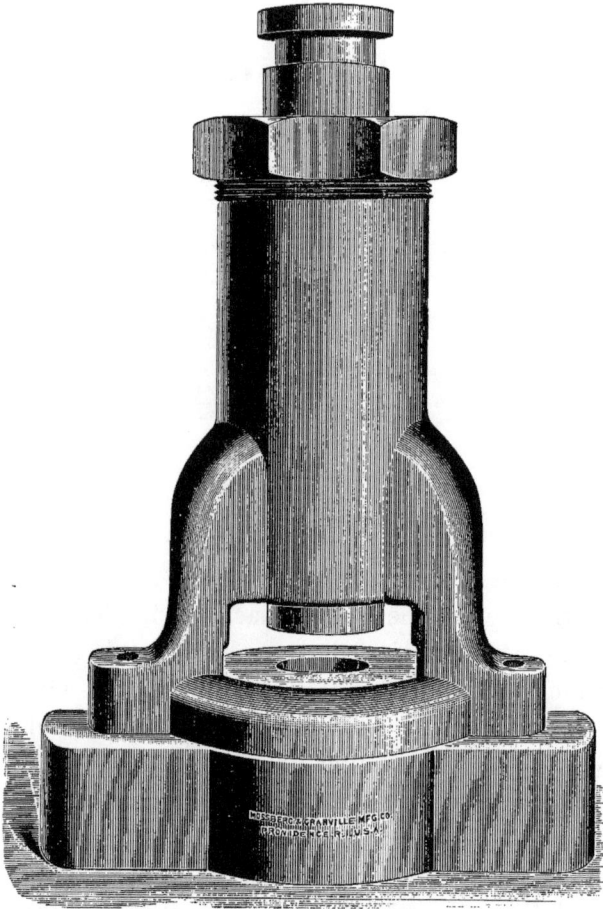

Fig. 40.

A sub press, is, as the name indicates, a press set under a press, and was designed by Mr. A. L. Dennison of Waltham, Mass., for the making of watch movements, and at first consisted of two pins acting as guides for the moving part of the sub press, which being quite a distance from the punch, served to make it enter the die correctly.

A very interesting article on the origin of the sub press, as well as the illustration and description of the first die thus made, by Mr. Dennison, can be found in the *American Machinist* of

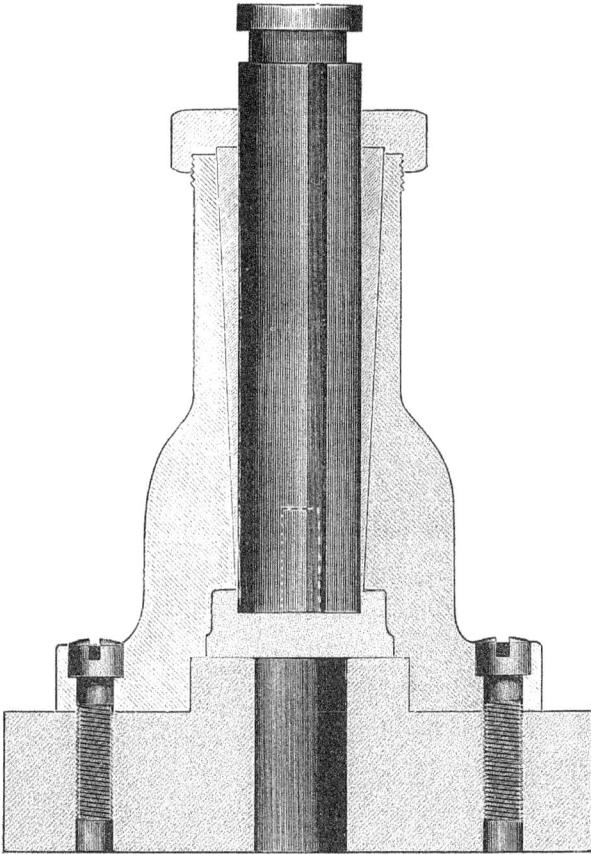

Fig. 41.

March 5th, **1891**, and will repay anyone for looking it up who is interested in this subject.

For regular sub-pressed die work, the form shown in Figs. **40** and **41** will be found a good style to use. The inside of the casting, shown in cross section, is made tapering, and bored out true with the plunger, and a heavy nut is fitted to the top, as shown in the illustration, so that the babbitt with which the sub press is filled is kept tight by screwing down the nut and forcing the babbitt into the taper and against the plunger.

In making a die for the sub press, it is made and fitted in the usual manner and set in its place. The punch is then carefully

set, and the space around the plunger is filled with the best quality
of babbitt metal.   The plunger has three parallel grooves milled
to prevent the liabilty of its turning in use, and the upper half or
arch of the press is shouldered on to the base and then held to the
same by screws, shown in the cross section Fig. 41.   The plunger
is driven by the slot shown at the top, and is attached loosely to
the gate of the press.   No dependence whatever is placed on the
ways of the press, but the stiffer the press in which the die is run
the better the results obtained, as it is considered good practice
that the punch should never enter the die, but should come down
and punch out the work without the edges coming in contact.

The illustration on page 34, shows how a die may be fitted up
when the height of the die space in the press will not admit of a
regular sub-press casting being used.   The punch and die are made
in the regular way, and after being finished the punch is entered
into the die about $\frac{1}{16}''$ and is kept in position by parallels being
placed between the two.   They are firmly clamped together, and
while thus clamped the holes are drilled, reamed and the four
pins, only two of which are shown in the illustration (which are
hardened and ground), are fitted in the holes in the bed of the
lower die, then the clamps and the parallels are removed and the
die is ready for use.

These pins insure the punch entering the die perfectly, and do
not depend on the ways of the press or the skill of the operator in
setting the same.

The advantage of having a die sub pressed, especially if it is a
complicated one, can hardly be over estimated, as a die which
before being sub pressed would cut but two to four thousand
blanks with the ways of the press as carefully adjusted as it was
possible cut twelve to forty thousand after it had been thus
arranged, and the work was of a superior quality.

SOFT PUNCHES AND DIES.

When punching thin stock, sheet iron, or soft metal such as copper, brass, etc., a soft punch will give equally good results to a hardened one, for when worn it can be taken out and upset and then "sheared in" by putting it in the press and forcing it through the die. When the blank or the piece punched out is the object sought, as in playing cards or cards for photograph work, the reverse of this plan is followed. The punch is made straight, hardened, the die is left soft as it gets worn, is set in with a pean of the hammer and the punch forced through it, thus practically making a new die, also keeping it exact to the size wanted for the work. Where the stock is very thin, machine steel will be found equal if not better than tool steel, and the operation of upsetting the punch will harden it, so that it will become very tough.

A punch of this kind well fitted will run off eight to sixteen thousand blanks from soft metal, and with less injury to the die than though a hard punch was used. For forming up sheet tin or other kitchen ware, cast-iron dies and a lead force are often used; the die being set in the press and a punch shank is secured in the hammer and then the melted lead poured around the shank, filling the die and forming a lead force. This is an ancient practice and is seldom followed in an up-to-date shop.

In exceptional cases the punch and dies can both be left soft to good advantage. This is particularly the case when the work to be punched is very thin or soft metal, or an intricate pattern that would be liable to crack or change in hardening; or if they are to be used on paper, which if cut in thin sheets is a very easy material to work, but when large numbers are punched at the same time it is one of the most obstinate of substances, on account of the toughness of the same.

A most interesting article on the punching of paper or fibre, which is practically the same thing, appeared some months ago in the columns of the *American Machinist*, by Mr. Oberlin Smith, who has made some most interesting tests on the power and tools required. Although not classed as soft punches, fluid punches are often used in expanding or distending work described on page 36.

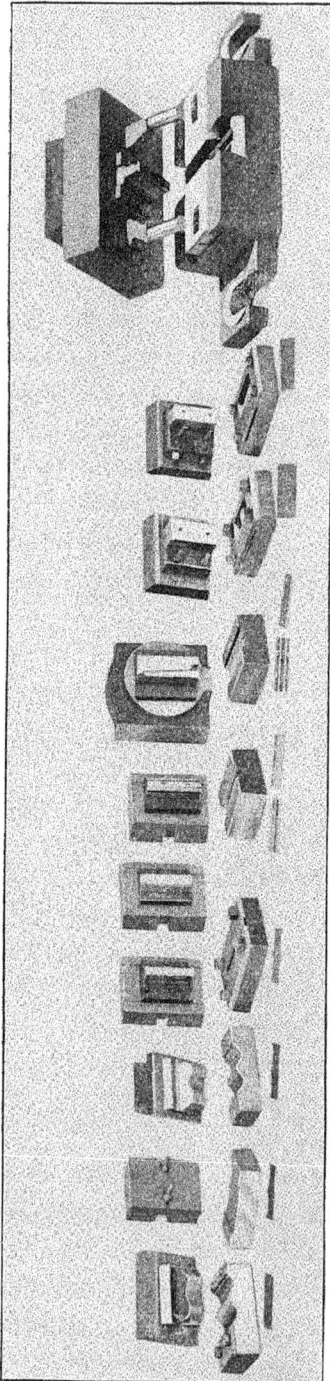

Fig. 42.

A SET OF DIES.

The illustration shows a set of dies designed for blanking, forming and piercing a set of springs and holders intended for holding cartridges in rapid firing army rifles.

The second and third dies are plain gang dies designed for blanking and piercing the clip.

The large die shown at the right is for bending up one of the clips at one operation, and is constructed on the same principle as the one shown on page 46, the side benders completing the springs at one stroke of the press.

The fourth and fifth dies, and the three punches over them, are for bending the second spring, three operations being required.

The fourth and last dies are for bending and piercing the springs, one of which requires three operations, and the other but one, the whole being shown to illustrate the number of tools often required for very small articles.

An attempt has been made by the photograper to illustrate the work done, by placing a spring in front of each die, with, however, but partial success.

CHUCK FOR MAKING PUNCHES.

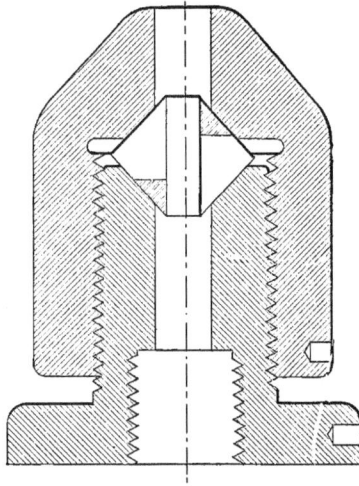

Fig. 43.

The advantage of having a chuck for turning out punches from solid stock, capable of taking a heavy chip without springing and still run perfectly true, is well known.

The illustration shown, represents one which will fill all wants in that direction. It is simple, and can be made in the shop, no special tools being required. The boys in the shop used to call it a "Pratt & Whitney" chuck, because a part of the tool, at least, is similar to one which is largely used by that firm, but the split cone is, I think, original. The base or internal portion of the chuck is screwed on to the mandrel of the lathe in the usual manner, and the split cone, which is provided with bushings to accommodate all sizes of steel, from one inch down, is gripped by the outer shell, which is worked by a spanner similar to the ordinary chuck.

For those who have much of this class of work to do, they will be well repaid for the trouble and expense required to make them, for made in the shop, the cost will not exceed $15.00 each. If some enterprising firm would take hold of the matter, they probably could be produced for half this sum.

SPRING DIES.

Fig. 44.

The half-tones show a combination punch and die for cutting and bending a sheet metal spring, the stock being fed in with a roll feed, and then removed by the use of an inclined press, shown on the opposite page. Any length of spring, up to the length of the die, is made by simply changing the length of feed. The stock is cut off and fed by gravity to the bending die, and after being bent, is removed in the same manner, the bending up of the ends of the spring allowing it to pass freely from the die.

Considerable trouble was experienced in getting the die to work nicely, but it was accomplished, finally, by the use of kerosene as a lubricant on the under die.

The press shown is what is known as a "Style B" Press, and the illustration is loaned by the courtesy of the Mossberg & Granville Mfg. Co., by whom the dies shown were made. The same style of press is largely used on can, pail, and other tinware, in connection with a single action die shown on page 25.

Fig. 45.

INCLINED POWER PRESS

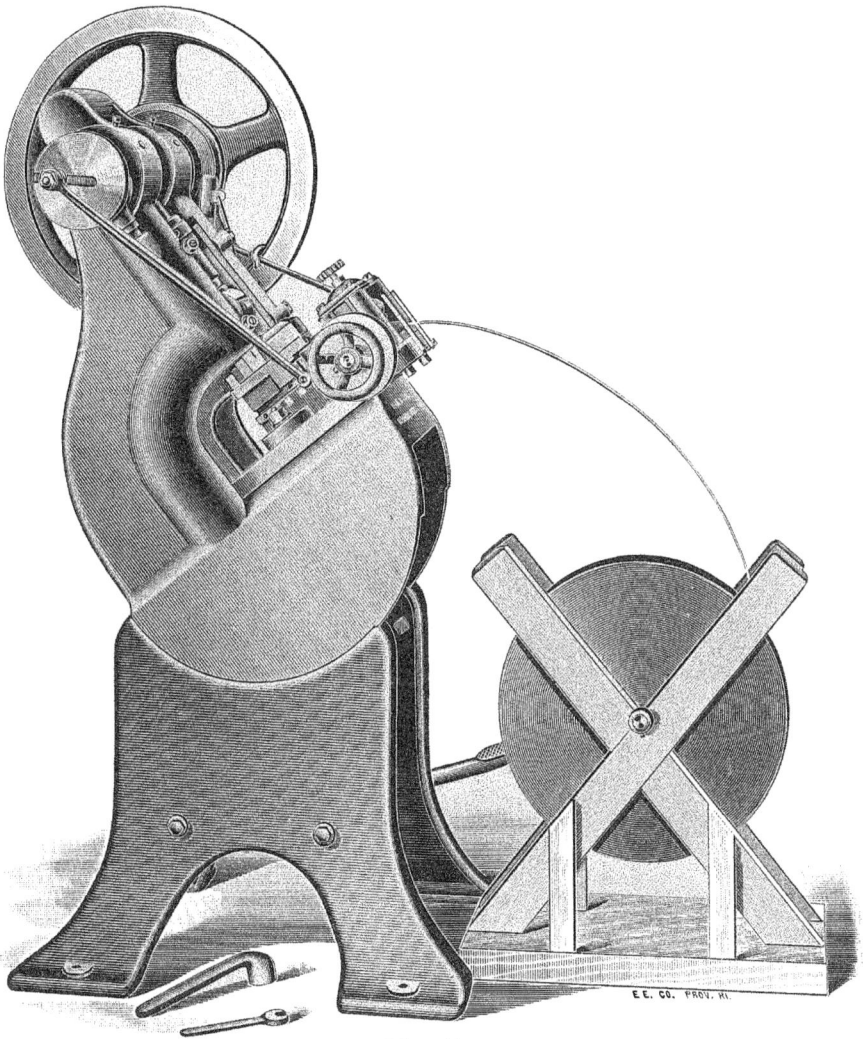

Fig. 46.

The advantage of a press of this kind is that single action combination dies, shown on page 25, can be used either in connection with a roll feed or without, and the work removed without any special attachment; the operation of the die being that, as the punch rises, the work is raised to the top of the die and removed by gravity. The expense of the die is greater, but a larger amount of work can be accomplished than with the "Push Through" double action die shown on page 27.

The greatest disadvantage of the inclined press is the inconvenient position of the press for die setting. This is avoided in the style shown, which can be instantly changed to the upright position.

DIE HOLDERS AND BOLSTERS.

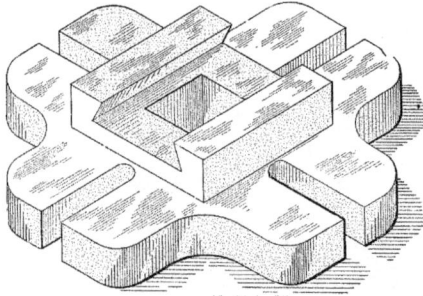

Fig. 47.

Die holders and bolsters are used for holding dies to the bed of the press, the holder often being used to support some part of the die that would otherwise be unable to stand the strain of punching. They are usually made of cast-iron, except for heavy work, when steel casting will be found preferable, as steel does not crack under any strain, and will stand more abuse than the cast-iron holder.

The holder shown in Fig. 47 is a good style, and is adapted to a large variety of work. It can be set either way in the press, and with either keys or set screws for holding the dies. If screws are used, the dies should be made of standard width, as 2″, 4″ and 6″, and thus avoid a large variety of holders. Dies for drop presses are usually held in a round block, as they can be more readily set in the drop, and are held by four or six screws, known as "poppets," as shown in Fig. 82 on page 84. This manner of holding a die will give good results, and it is suitable for all drop press work, except drop forgings, for which the dovetail slot in both hammer and anvil will be found preferable.

The four holders shown in the large cut, Fig. 48, are intended for taking a uniform range of dies, and the shoe is so designed that it may be removed and replaced by simply loosening the bolt, as it is not necessary to remove the nut, the slot being so designed that the holders may be pulled foward, and then sideways, and thus removed from the press.

The bolster plate, shown in the last illustration, is intended to be placed on the bed of the press, and the die holder fastened by the use of two cap screws or stud bolts, which are not shown in the illustration. The die holder and bolster are used as shown on the press on page 59.

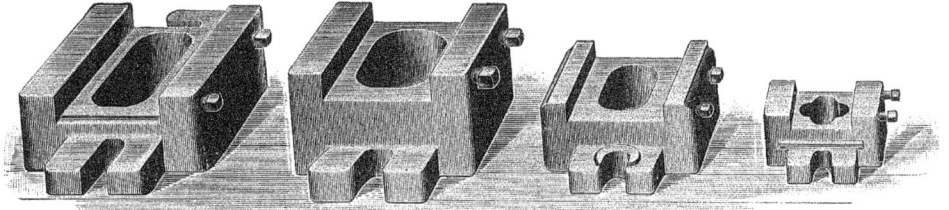

Fig. 48.

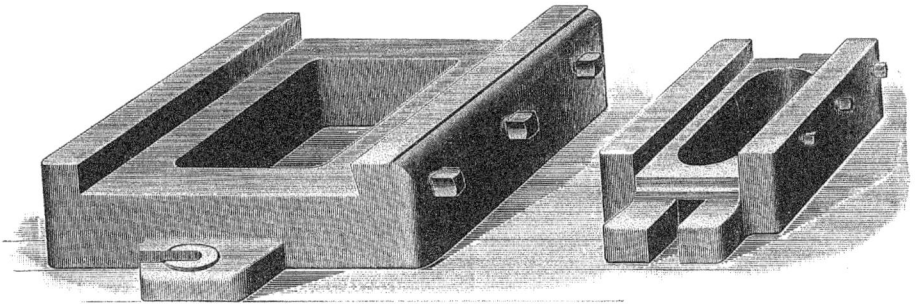

Fig. 49.

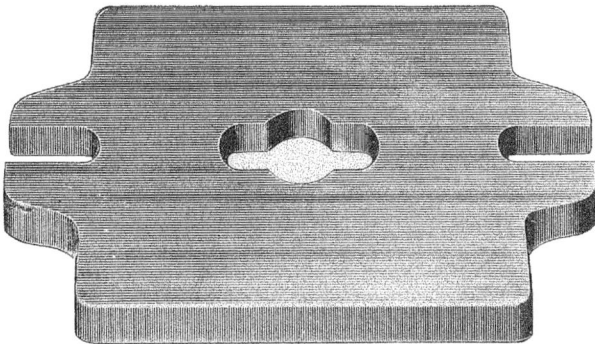

Fig. 50.

A GANG PRESS.

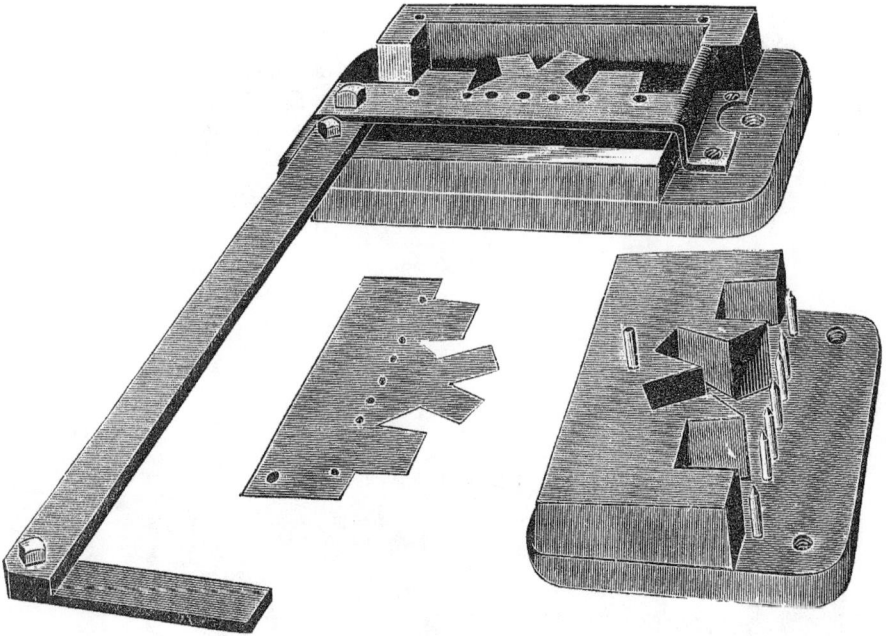

Fig. 51.

The illustration on the opposite page, Fig. 52, shows a gang of foot presses; Fig. 51, one of the dies used in the same, and intended for notching and mitering skylight bars. They are made with three, four or six presses on one table, and intended to be used in consecutive order, each press being fitted with dies for performing a certain operation on the bar.

The die and punch, which are shown with the stripper removed, are so arranged that the back of the die acts as a guide and prevents the punches from sheering and being forced away from the die when cutting.

The press is made by the West Manufacturing Company of Buffalo, N. Y., through whose courtesy the illustration of the same is shown.

A GANG PRESS.

Fig. 52.

CAM STRIPPERS.

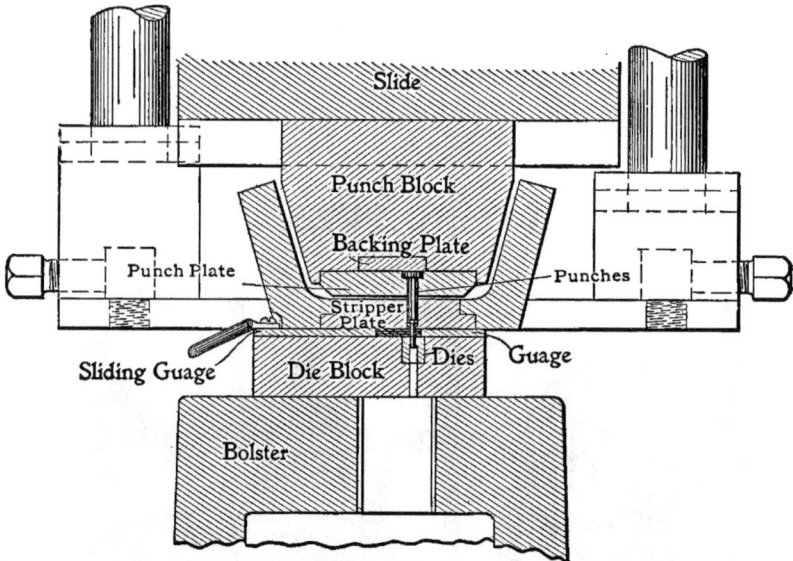

Fig. 53.

Figs. 53 and 54 show a die having a cam-actuated stripper, also press specially adapted to the use of the same, illustrations of which are loaned through the courtesy of the E. W. Bliss Co.

The great advantage of a stripper of this kind is less liability of breaking the punches, and the thicker stock that can be punched on account of the decreased length of punches that can be used, which are made very short, and are held in a punch back. The stripper plate is fitted with bushings, which guide the punches and support them while piercing the metal.

Fig. 53 shows the details of the die and punches, the punches being held in the punch block which is fitted with a backing plate of hardened steel; the stripper plate is attached to four vertical rods which receive their motion from two cams on the main shaft. When the stripper is up, a free space is left between punch and die, enabling the operator to manipulate his work quickly and accurately. The stripper comes down on the blank, straightening it and clamping it tight before the punches enter, and holds it while the punching and stripping is done, and the work comes out perfectly straight and true. As the stripper moves up and down with the punches, they may be made shorter than a stationary stripper would permit, and the punches used are therefore more rigid and lasting. This also enables the punching of smaller holes in proportion to the thickness of the stock, the stripper acting as a guide for the punches up to the moment when they enter the metal.

CAM=ACTUATED STRIPPER PRESS.

Fig. 54.

The press shown has a 6″ crank shaft, weighs about 15,000 pounds, and is geared 7½ to 1, and is one of the latest productions of the E. W. Bliss Co.

U. S. STANDARD PUNCHES AND DIES.

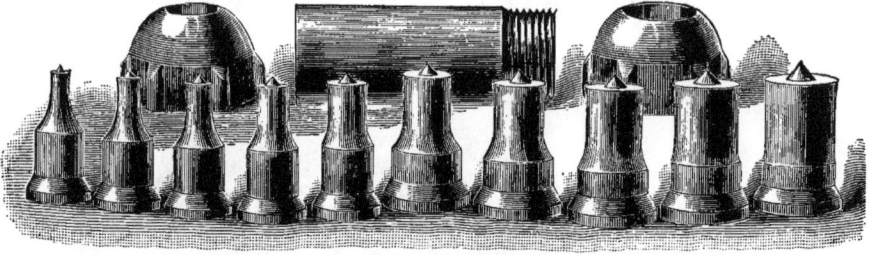

Fig. 55.

The following illustrations are loaned through the the courtesy of Mr. I. P. Richards, the well known inventor of the only standard for round punches and dies, and the best method for holding the same. The shank which is shown in Fig. 57 is fitted to the press, and the punches are held in place by the coupling. The punches being a standard length, can be easily renewed when worn, or exchanged for a larger or smaller size without disturbing the adjustment of the press. They are made in all sizes from $\frac{1}{8}''$ to $4''$ diameter, and adapted to almost all classes of round work, but more especially for punching out holes for boiler, tank, and other work which is secured by riveting.

The punch shown in Fig. 58 is ground shearing thus reducing the power required for operating the same. These punches are furnished by the inventor, Mr. Richards, at a less cost than they can be made by consumers, although the patent on the same having expired, they are open to the public, and are extensively copied by other manufacturers, often, we are sorry to say, without giving Mr. Richards the credit which is due for the invention and his efforts to introduce and maintain a standard punch.

While we have no data of the power required to operate these punches, we have reason to believe the statement that 40% is saved by their use.

Fig. 56.

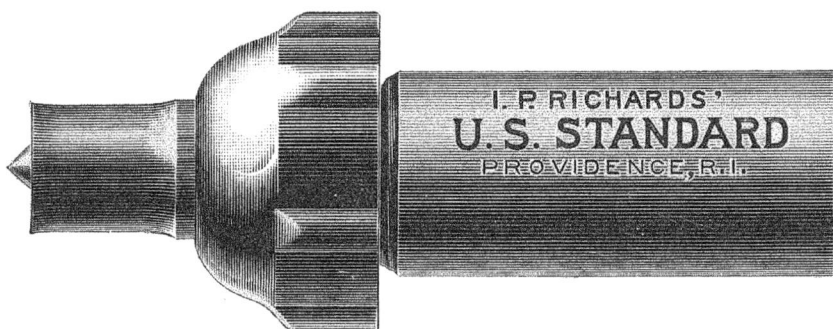

I. P. RICHARDS'
U. S. STANDARD
PROVIDENCE, R. I.

Fig. 57.

No 7

Fig. 58.

PUNCHES AND DIES FOR NUT WORK.

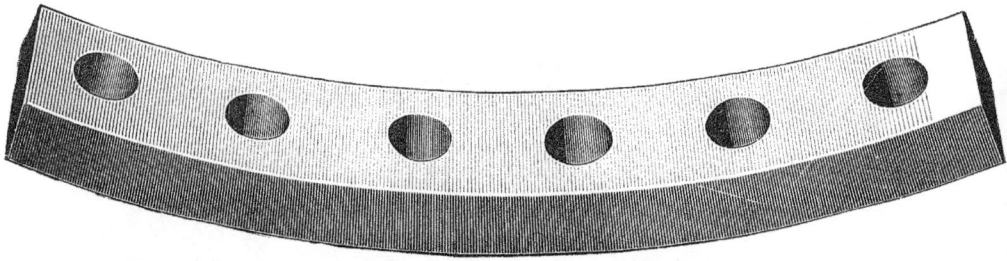

Fig. 59.

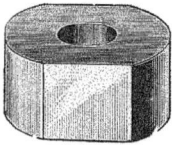 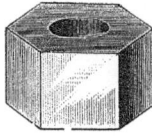 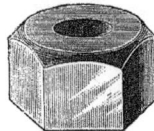 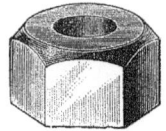

Fig. 60.   Fig. 61.   Fig. 62.   Fig. 63.

In the manufacture of nuts it is customary to make each operation a separate one; that is, the bar is first cut up into short lengths, and then the holes punched the entire length of the bar; the nut is then roughly blanked out, and put through, or rather under, a crowning punch which presses the corners, and leaves the circle on the top, shown in Fig. 62. The nut is then put through one, two, or three shaving dies, each of which takes off a thin shaving, and leaves the side of the nut very smooth. The greater the number of operations the nut passes through, the better the finish obtained.

Various attempts have been made to reduce the number of operations required in this work, and thus produce a nut at a less cost. These attempts have met with more or less success. Perhaps the nearest approach in this direction has been made by the Port Chester Nut Co., who punch a blank and shave a nut at each stroke of the press. This is a patented process, and for that reason is not shown here.

The Figs. 59, 60, 61, 62 and 63 show the first, second, third, fourth and fifth processes. The curvature of the bar is caused by the first punching, as the metal invariably stretches more on the under side; this however is but of little consequence as it does not show in the finished nut.

FORGING BICYCLE SPROCKETS.

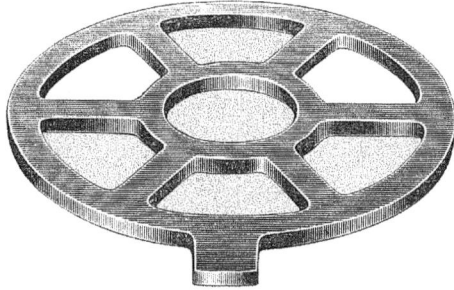

Fig. 64.

Fig. 64 illustrates a method of making bicycle sprockets that is both better and cheaper than that usually followed, and one that is not in general use. It is the invention of Mr. L. B. Munger,—who is well known for his improved methods of manufacturing bicycle parts,—and has been copied to some extent by his competitors.

The sprocket is blanked out in the usual manner, and the cross area of the stock that is left to form the spoke is carefully calculated, so that after the sprocket has been through the forging die, no fin is left around the edges of either spoke or hub. This calls for good judgment as well as accurate measurement on the part of the die-maker, as it has proved a failure in several cases where an attempt has been made to adopt Mr. Munger's method.

The tail shown on the blank is intended to be grasped in the tongs while heating, and in placing the blank upon the forging dies, and cut off after that operation.

This same method is followed by Mr. Munger in all kinds of forgings the work being first punched out to the proper size, leaving the required amount of stock necessary to fill the die, and is then forged, leaving little or no fin, and avoiding the trimming process after the work has been forged.

This method has been jealously guarded by Mr. Munger, and is here published for the first time with his full permission.

A GROUP OF ARMATURE PUNCHINGS.

12 IN.

Fig. 65.

Illustration Fig. 65 shows a group of armature punchings, all of which are blanked out by means of compound dies, the largest of the group being 36″ diameter, and the smallest one, which is shown in the centre, 1⅛″ diameter; the others being respectively 4″, 6¼″ and 20″ diameter.

A detailed description of the dies for producing this work is given on page 32. The methods of constructing the dies are so varied that it is impossible to give a plan which will fit all styles of armature sheets. Plans or sketches for dies for any work of this character will be furnished on request.

A GROUP OF SMALL PUNCHINGS.

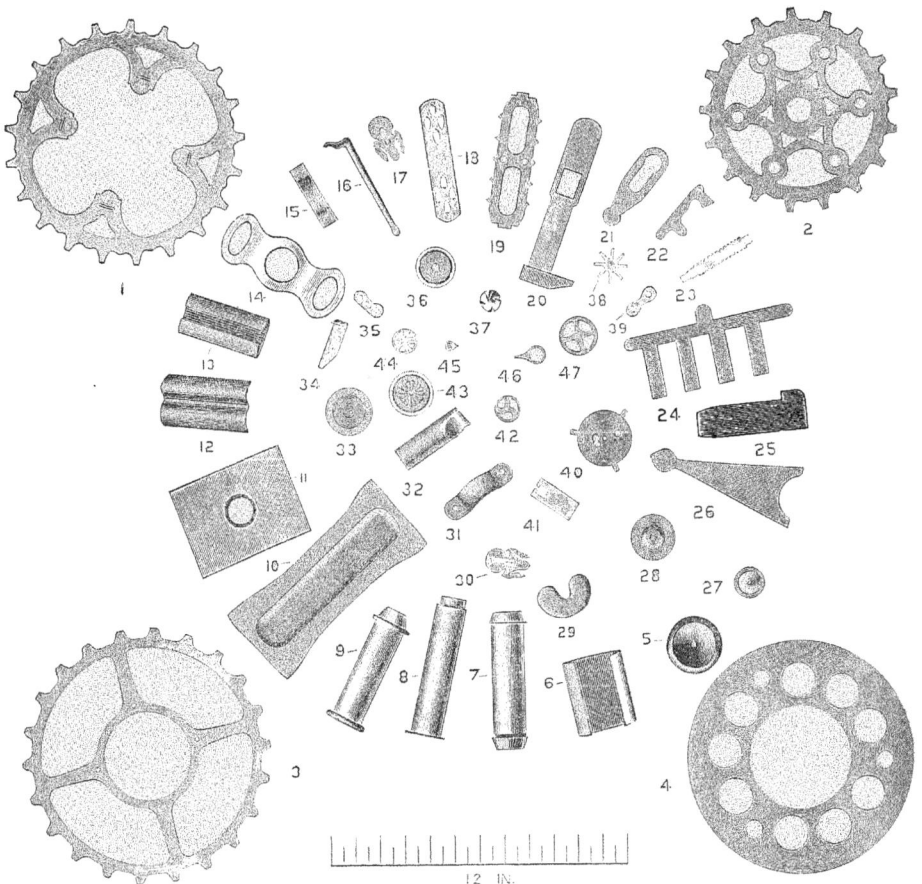

Fig. 66.

The above illustration shows a large variety of different articles which are produced wholly, or in part, by press work.

The four corner illustrations will be recognized by those interested in bicycle manufacture, being punchings of sprocket wheels made by four of the leading manufacturers. The figures 14, 15, 19, 21, 31 and 35, are also parts for bicycles; and numbers 32, 42, 44 and 47 represent the pieces used in the construction of the well-known bicycle whistle.

SPEED OF PRESSES.

The speed at which a press can be operated depends upon the nature and thickness of the stock being punched, the quality of the steel of which the tools are made, and the material used for lubricating the same. In punching a soft material, such as paper, cloth or very thin sheet metal, a speed of 200 to 350 revolutions per minute is practical, as the fiber of the metal does not require to be broken, as is the case when operating on thick or hard material.

The old-fashioned rule that a punch will penetrate metal equal in thickness to the diameter of itself, is fast being exploded, as it is a well known fact that a $\frac{1}{2}''$ punch can be put through a $2''$ bar of iron, provided the tools are properly made, and a speed of not over one stroke in two and one-half minutes be given to the same; breaking the punch being the result of an increase of this speed.

In ordinary punching sheet metal, .014'' to .04'' thick, 125 to 150 revolutions per minute has been found as high a speed as it is practical to run; in blanking out large sheet-iron armature disks, .025'' thick, and 20'' diameter, more work is produced at 60 revolutions than at 90, as the dies will stand up to the work and run off a larger number of sheets without being ground, than when used at higher speed.

The speed of drawing punches depends largely on the work done, and the material used for the same, as a slow speed will often enable work to be produced that would rupture at a high speed. On large and deep shells, a speed of two to four revolutions is often necessary, while small and shallow work will stand a speed of 150 to 200 strokes per minute.

The stroke of the press is an important factor which must be taken into consideration when estimating the speed at which work may be punched, the idea being that the tools must have time to break or rupture the grain of the metal, and this can be facilitated by giving the proper amount of clearance to the punch, as well as the usual clearance to the die; also care being taken when punching a thick metal that the punch and die do not fit too closely. Probably more trouble is caused by this one defect in tool making than any other. It is hard to convince a good die-maker that the punch and die must not fit nicely when used on thick work; often a difference of $\frac{1}{32}''$ to $\frac{1}{16}''$ should be given on work $\frac{3}{8}''$ in thickness or over. On thin sheet metal the tools must fit nicely in order to produce the work without burr, and the speed of the same can be greatly increased; 90,000 to 100,000 per day being the usual output of a press running at 200 revolution per minute, the stock being fed with either roll or ratchet feed.

DIE SETTING.

Setting punches and dies in the press is often done without due regard for the tools,—that they are properly set and adjusted for the work they have to do. From a mistaken idea of economy, this work is often put into the hands of an operator who has little idea of the value of the tools which he is adjusting, and to whom any accident caused by improper setting is a small matter.

To set a punch and die in the press, remove the belt from the wheel, unless the press is provided,—as all presses should be,— with a countershaft. Place the die and holder, (into which the die can be keyed to better advantage on the bench), on the bolster plate, and enter the punch carefully by hand; now bring the gate or slide of the press down until the bottom of the slide strikes the shoulder of the the punch and prevents it from being moved out of the die; next ease up slightly on the slide, so that the punch, die and holder can be moved along by hand into their proper places in the press. The punch, if secured in the head block or chuck of the press, is pushed to the left hand side of the dovetailed slot in the slide, and the key driven in. The bolt or stud holding down the die-holder should now be slightly tightened, and a few raps with the hammer be given to the die; this causes it to find its proper place on the bolster; the holder is now secured to the bolster plate, and the punch withdrawn from the die and then entered again, to be sure that the operation of setting has been properly performed.

No specified distance can be given as to the distance the punch should enter the die. In sub-press dies for watch work, the punch should come to the surface of the die and enter the least possible amount; while in rough crude dies used for thick sheet iron or boiler plates, the removal of the scrap or the piece punched out will be greatly facilitated if the punch enter $\frac{3}{8}''$ or more.

The above method will not meet with the approval of many,

but a large experience in different methods of die setting has led to the description just given.

There is one good rule which, if followed, would practically prevent the accidents now so common in large press rooms, and that is: never allow the hand or fingers to be placed under the punch while the press is in motion. Many of the accidents which occur are caused by gross carelessness on the part of the operator who is induced to take the chances of accidents because of the larger amount of work which can be produced.

BROACHING.

Fig. 67.

Although not usually classed with punches and dies, the operation of broaching is analagous with punching, and for that reason is properly entitled to description here. The broach is really a punch, the cored or drilled hole in the work broached forming a die for the same. It is used for finishing holes which may be either punched, cored or drilled in brass, iron or steel, and which may be round, square or any other desired shape. The operation is best performed in a press designed for the purpose, having an adjustable stroke of $1''$ to $12''$. The broach shown is intended for finishing a cored hole, and is $2\frac{1}{2}'' \times \frac{1}{2}''$ rectangular section, about $8''$ in length. The teeth being very coarse on the lower end, are intended for taking out the bulk of the stock, until the centre of the broach is reached, when they are sheared in opposite directions, thus breaking the chip off, and then the teeth decrease in size until the upper end of the broach is reached, when the same is left perfectly straight for about $2''$, forming a sizer, keeping the finished hole a standard size. The broach is best driven when operated on by a V block secured in the press in the same manner that the punch is usually held, and as the plunger comes down, the broach will find its own centre with less liability of breaking or making an imperfect hole. Broaching of almost any length can be done in a press having a short stroke, by using a successive number of blocks, the broach being inserted in the hole, and driven down the full stroke of the press. A block of the same thickness as the stroke of the press is then inserted, and the operation can be repeated until the desired length of drive is obtained. The results obtained are not equal to a continuous stroke broaching press, as the stopping of the broach when partly through the work, gives the metal an opportunity to settle into the teeth of the broach, thus increasing the liability of breakage.

There are on the market several machines designed expressly for this work, several of which perform the operation by pulling the broach through the hole, which is an improvement over the method just described, but as it does not come under the jurisdiction of this work, the description of this machine is omitted here.

REVOLUTION COUNTERS FOR PRESS USE.

Fig. 68.

The use of a revolution counter or indicator for keeping account of the work produced by a press, is preferable to the ordinary manner of weighing up or trusting to the intelligence of the operator to determine the amount of work done. This fact has led it to be adopted by several of the largest press rooms in the country, and with satisfactory results.

The above illustration shows a style which is adapted to all makes of presses, as it can be connected either directly to the shaft itself, or to the centre, or can be located in front, on the upper part of the press, and then connected by a lever from the slide, and will give an accurate record of the revolutions made. It is simple, having no springs or gears, and can be furnished with either four, five or six dials. The five figure indicator shown in the illustration will count up to 99,999, and then repeat auto-matically. A dial can be set to zero, or the counter can be had without the thumbscrews for re-setting.

In using the counter in connection with gang punches, that is, in work that requires one are more strokes of the press to properly start the strip of metal used in feeding, there is a loss of six to nine per cent. from the number actually indicated. This fact being thoroughly understood, the proper amount of work can be produced, making the necessary allowance for the false strokes.

BALL CHUCK.

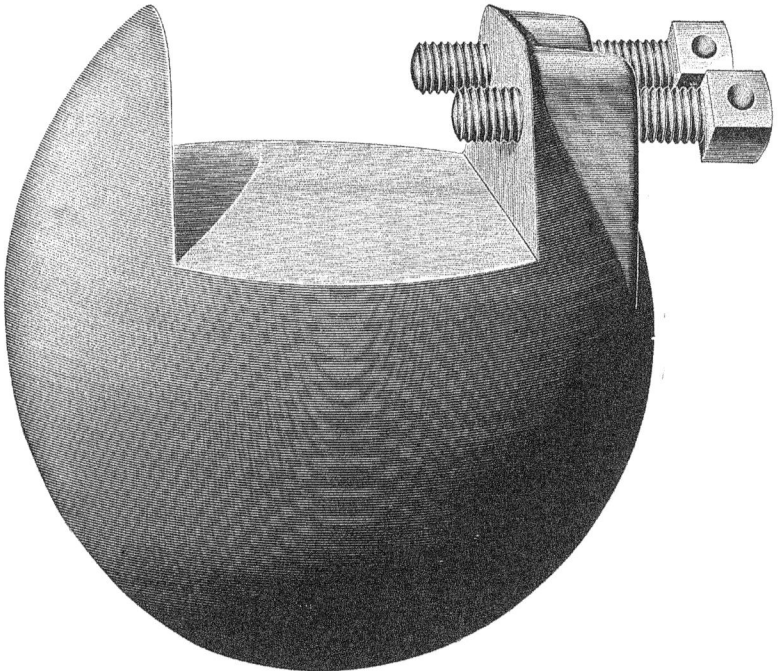

Fig. 69.

The illustration Fig. 69 shows a chuck for bench use, which is adapted for "working out" punches and dies or any work which can be done more conveniently by placing it at an angle, so as to use a hammer and chisel to good advantage. The chuck was originally designed for and is usually used in die sinking, but the advantages mentioned has led to its use in the die room, and once introduced, it will be found to be a good tool.

A base provided with a cavity of the same radius as the bottom of the ball is intended to be used in connection with the chuck, and is usually made by coiling up an old belt, and then striking the chuck firmly down into the same. The die is held in the slot, secured by the two set-screws.

The advantages obtained from the use of this chuck will be appreciated by those who have experienced the difficulties of chipping out an odd-shaped die.

FINISHING PRESS WORK.

Fig. 70.

Perhaps a few words on finishing punched work may be of interest to those who have the good or bad fortune to have charge of that kind of work, and the barrel shown is specially adapted to work of this kind; being a modification of the ordinary tumbling barrel, so arranged that the incline may be varied, giving a greater or less effect on the work as may be wanted.

All punched work as it comes from the press is in an unfinished condition, owing to the soap, oil or other substance which is used for lubricating the dies, and the advantage of tumbling cannot be over-estimated. Burrs can be removed with practically no expense, and work put in a salable condition which would otherwise be consigned to the scrap.

For abrasive materials to be used in tumbling the work, there is nothing better than broken emery wheels for rough tumbling or taking off heavy burrs, and for finishing light, delicate work, leather chips and sawdust will give good results. For polishing buttons, of hard rubber or papier mache, old woolen rags will be found to give a high polish without marring the work.

For tumbling shells which have been drawn, but are to be put through a second operation, a solution of strong soap water, either hot or cold, will be found to give a good polish and also leave the work in good condition for the second drawing.

BELTING UP PRESSES.

To belt up a power press properly, set it in line with the countershaft or main shafting, and fasten securely to the floor, taking care that it is properly leveled, and that each leg has an equal bearing on the floor. The better the foundation, the better results will be obtained. Take double the distance from the centre of counter or main shaft to centre of crank shaft on the press, and add one-half the circumference of both pulley and balance wheel, and the result will be the length of belt required. If the press is an inclined one, and it is to be used in an upright position also, set the press on an incline and cut the belt to the proper length; then set it upright and set a piece of belt of the proper length to be used when the press is in that position. Put on a belt the full width of the wheel. Do not put a 2″ belt on a 4″ balance wheel, and then expect the momentum of the wheel to do the rest. Do not set the press directly under the main or countershaft when it is possible to avoid it. Set it at least 2 ft. away, even if you have to sacrifice floor space to do so. Run the belt with the grain side next to the pulley. A long belt is preferable to a short one, as the adhesion caused by the weight of the belt is more desirable than that obtained by tight lacing, and the belt will wear longer and give better results.

FEEDING PRESSES.

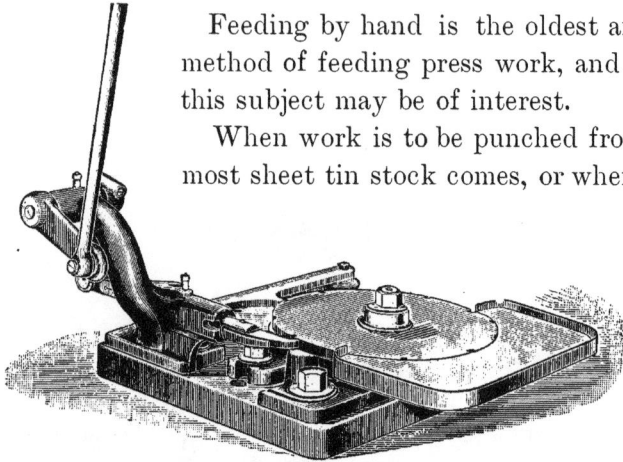

Feeding by hand is the oldest and most common method of feeding press work, and a few words on this subject may be of interest.

When work is to be punched from short stock, as most sheet tin stock comes, or where irregular stock is used, an automatic feed is of doubtful utility, for the operation of keeping the feed supplied is certainly as troublesome as feeding the

Fig. 71.

press itself. In cases of this kind it is best to feed by hand as a skillful operator moves his hands with the press and will produce work rapidly, but if the work will admit of it, a power feed should be used.

We describe a few of the many kinds in use: roll, finger, slide, double roll, reel, tube, grip, ratchet and friction dial feeds, combination feeds and knock-outs.

When large quantities of the same general class of work are wanted, and the stock can be procured in long strips or rolls, fit the press with a plain roll feed, Fig. 72, as they are easily applied, and can be quickly adjusted and changed from one class of work to another. They are made in several styles, to feed from front to back, left to right, or the reverse, single or double and with ratchet, friction, or "pinch" adjustments of rolls. The amount that can be fed is usually regulated by the diameter of the rolls except in special cases where a rack and pinion, giving any amount of feed wanted, can be used.

Fig. 72.

A finger feed is an adjustable "finger" fitted to and worked by the slide of the press, so adjusted that it is raised at each stroke just enough to allow the stock to pass under, and then catches in the hole punched by the preceding stroke. It is inexpensive, and is not in the way when not in use. When fitted with a power reel to wind up the scrap, it becomes one of the most useful feeds known, and can be used equally well on both single and double acting presses.

Tube feeds are used for feeding work that has been previously punched out by a plain die. It is best adapted for coining, swagging, etc. The work is caught in a tube as it falls from the blanking die, and is then transferred into a second tube placed on the press. An automatic moving slide delivers the lowest piece in the tube to the dies at the proper time. This feed is largely used on coins, medals, buttons, etc.

An automatic slide feed consists of a mechanical hand which grips the work and then slides it to the proper place on the dies, and is specially adapted to a fragile material, as cardboard, paper, etc., and is also one of the most accurate style of feed in use.

A grip or claw feed is so arranged as to enter the first hole in the stock just punched, and then draw it ahead the proper distance for the next blank. It is easy to adjust, and can be run at a very rapid speed.

Dial feeds, Fig. 71, are used to hold a blank on which it is desired to perform a second operation, as armature blanks, or to act as a carrier to present the blank in successive rotation to the punch, and in some cases to remove the work after the punch has performed its duty. They may be belt, ratchet, or friction driven, and may be a smooth disk as used for cartridge work, or fitted with sockets to hold and carry the work.

A combination feed is two or more feeds in one, as they are usually adapted for special operations, and can not be changed with the same ease as the roll or finger feed.

Knockouts are fitted to many presses to assist in removing the work from the dies. They are applied to either the slide or body of the press, and in the latter case are connected by a pitman to the end of the crank shaft, so that as the punch rises from the die, the work is forced out.

HARDENING DIES.

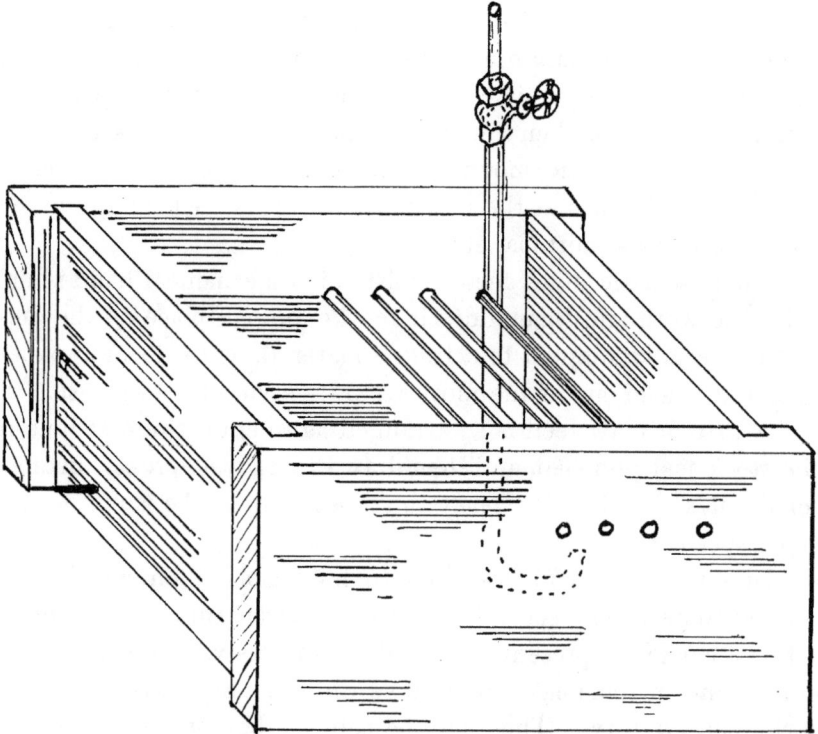

Fig. 73

So much has been said on the subject of hardening steel that we are disposed to let it alone, but feeling that a work of this kind would not be complete without it, will give the results of our experience. For heat we prefer a gas furnace of any good make, but this not available, a clean charcoal fire is nearly as good.

Fill all holes in the die that you do not wish to have a cutting edge, such as the stripper holes, with fire clay; use plenty of good hard wood charcoal, heat slowly but steadily to an even red. Dip the die entirely below the surface in salt water, for while there are baths that are good for special work, steel that will not harden in clean salted water is not fit to make a die of, keep it in the water until it is entirely cold, then put it on the fire again and warm it until the heat is as great as the hand can bear, then cool off and polish and draw to a light straw. The object of the second heating is to lessen the liability of fire cracking.

For large dies it is indispensable to have a tank properly arranged for cooling the steel quickly. This is best arranged by having two rods fixed across the tank about 10″ below the surface of the water; the water supply is led down below the surface nearly to the bottom of the tank. The end is bent up so that as the die is plunged into the tub the valve is opened and a circulation of water is created in the tank. This method will prevent the soft spots often met with in hardening. Fig. 73 shows the arrangement so clearly that no further explanation is required.

Fig. 74.

Although not properly classed as a punch and die, and yet clearly the result of the laws of evolution, the above illustrated tool is entitled to be represented here; and shows the trip hammer in its most primitive form. It is taken from a German illustration, loaned through the kindness of the Bradley Co., of Syracuse, N. Y., for insertion here.

The mode of operating the hammer is clearly shown in the illustration, the water-wheel attached to one end of the revolving drum carries the collar and dog for raising the hammer.

The method of keying the lower die in the anvil shown, is identical with that practiced at the present time, the force of the blow being controlled by the in-flow of water actuated by the knotted rope which hangs within easy reach of the smith. The swinging seat, however, would not be tolerated with the methods of operating the hammers now practiced.

The exact antiquity of the sketch could not be ascertained, but it is not far from the 16th century.

BROWN & SHARPE MFG. CO., Providence, R. I.
TOOLS FOR DIE WORK.

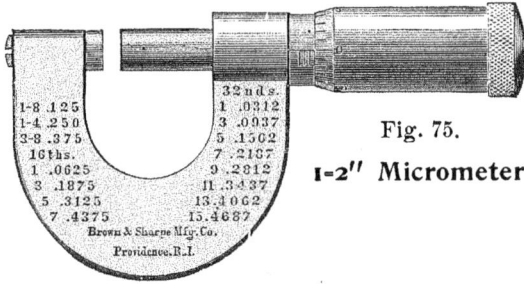

Fig. 75.

1-2″ Micrometer.

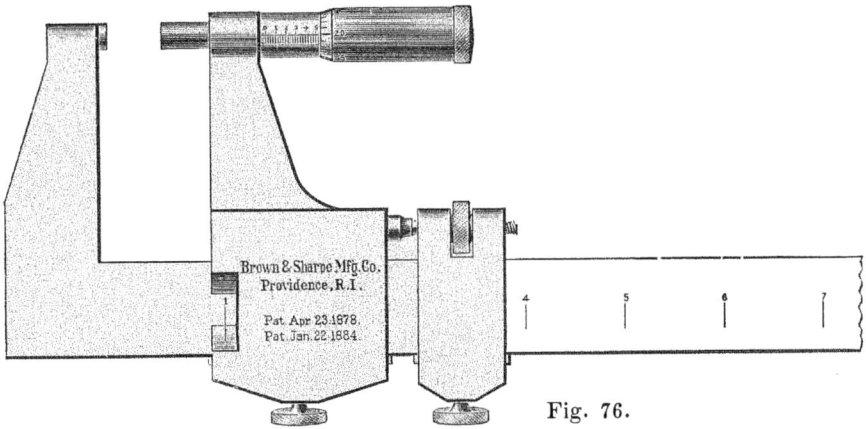

Fig. 76.

Micrometer Calipers.

Made to measure from ½″ to 24″, and to read to .0001″.

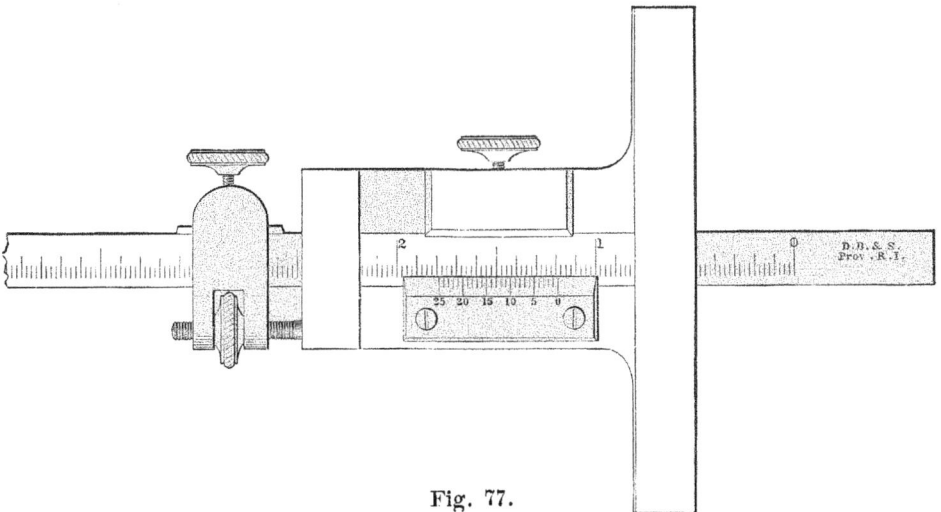

Fig. 77.
Height Gauge.

BROWN & SHARPE MFG. CO., Providence, R. I.

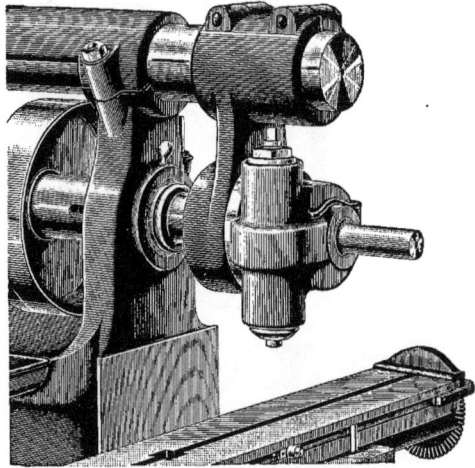

Fig. 78.
Vertical Spindle Milling Attachments.

Fig. 79

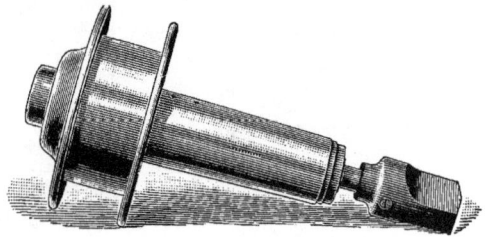

Fig. 80.

High Speed Milling Attachment and Driving Fixture.

Fig. 81.
Universal Vise.

Attachments used with Brown & Sharpe Universal Milling Machine for die work.

PAWTUCKET MFG. CO., Pawtucket, R. I.

Fig. 82.

Press for Blanking, Champering and Trimming Nuts.

Automatic Drop Press.

Designed for all classes of stamping and bending operations and can be operated by inexperienced labor without danger of injury.

MOSSBERG & GRANVILLE MFG. CO., Providence, R. I.

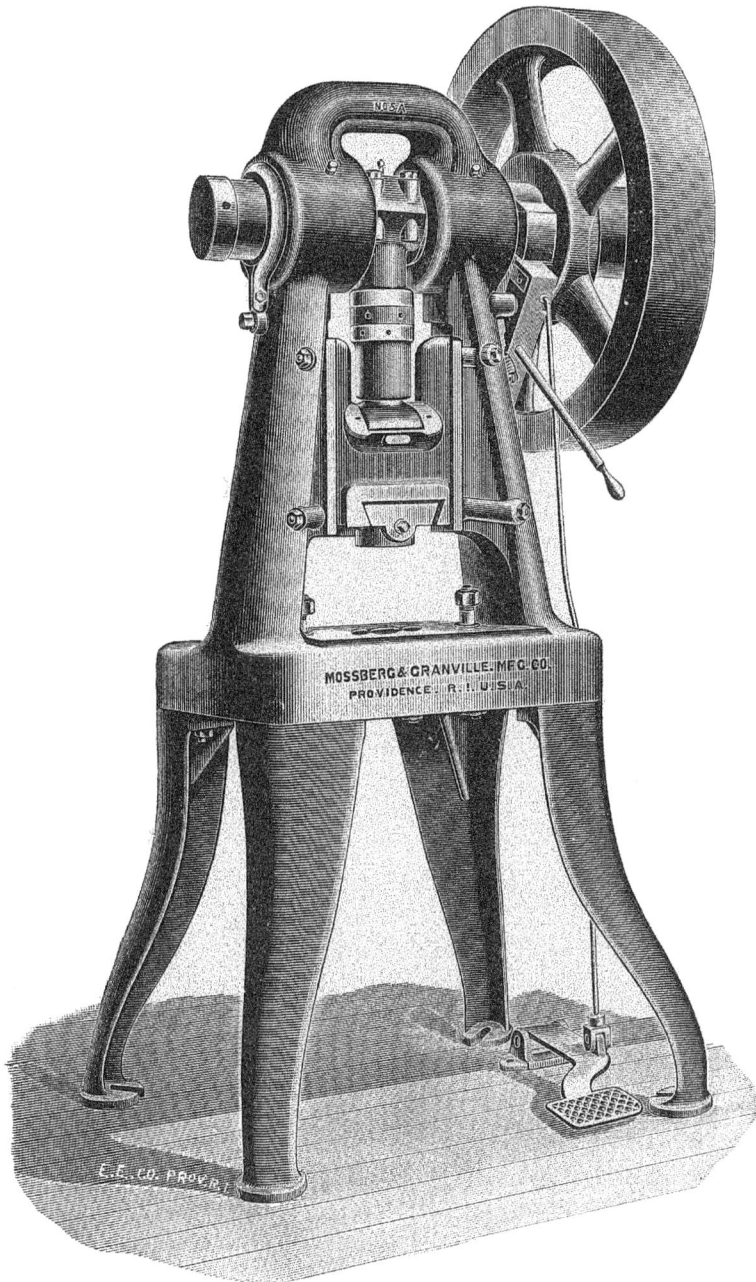

Single Acting Power Press.

Fitted with instantaneous clutch and stop motion. Designed for all classes of punching operations, and can be fitted with roll, finger or dial feeds.

E. W. BLISS CO., Brooklyn, N. Y.

PRESSES, DIES, SHEARS, DROP HAMMERS, DOUBLE SEAMERS, TRIMMERS.

Toggle Drawing Press.

The toggle drawing press as shown in above cut is used in the manufacture of wash bowls, milk and pudding pans, tea kettles bodies and breasts, sauce pans, bucket and pot covers, small scoops. cuspidors, coal hod bottoms, trays, dust pans, brass and copper goods.

This press is fitted with an automatic friction clutch; it can be stopped and started at any point of the stroke, by operating hand level not shown.

The toggle movement as applied to this press has greatly increased the durability and smoothness of action, insuring a more perfect "dwell" of the blank holder than can be maintained in presses of any other design.

E. W. BLISS CO., Brooklyn, N. Y.

EMBOSSING PRESSES, SHEET METAL WORKING MACHINERY, SPECIAL MACHINES.

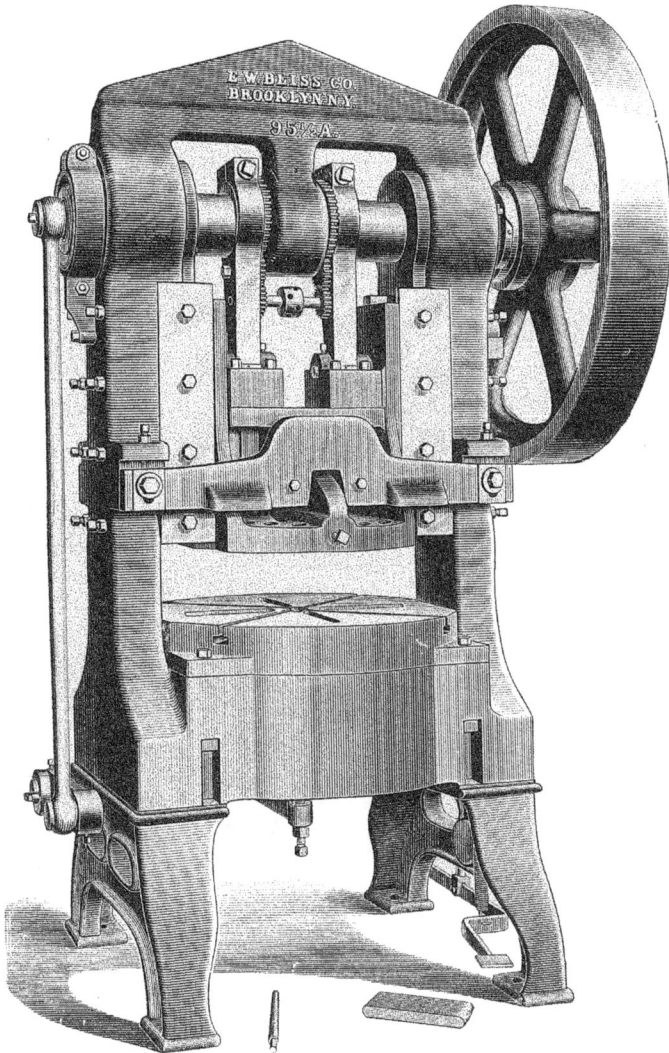

Armature Disk Cutting Press.

The press shown in above cut is specially designed for cutting the outside and inside of armature disks simultaneously, insuring exact concentric of outside and inside circumference.

This press is fitted with a positive knockout device.

The same general style of presses also supplied in larger and smaller patterns.

NEW DOTY MFG. CO., Janesville, Wis., U. S. A.

Above cut represents a combined punches and shears in which the knives for flat and round iron are independent of each other and of the punch; and both sets of knives as well as punch are at all times times ready for use without any change. While it is a single end, it has all the advantages of a double end machine, and is much less expensive.

E. S. STILES PRESS CO., Watertown, N. Y., U. S. A.

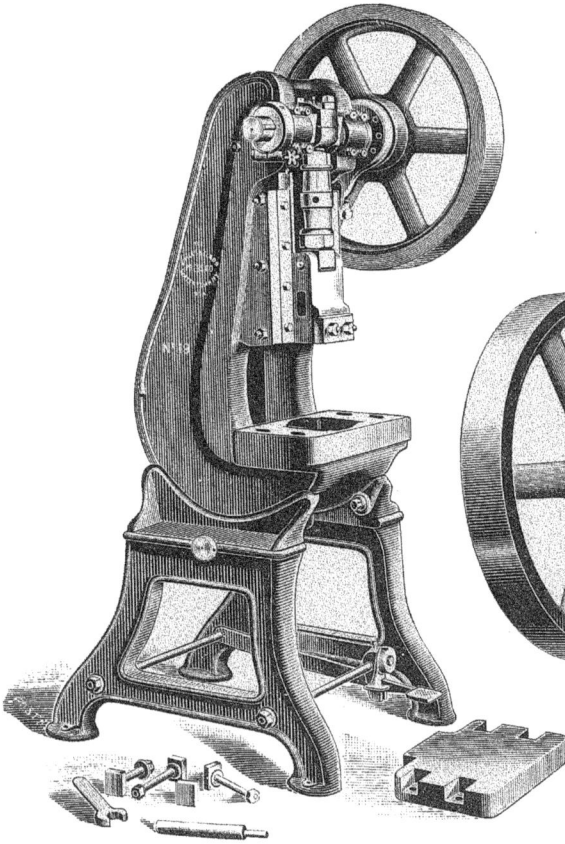

No. 19 Adjustable Incline Press.

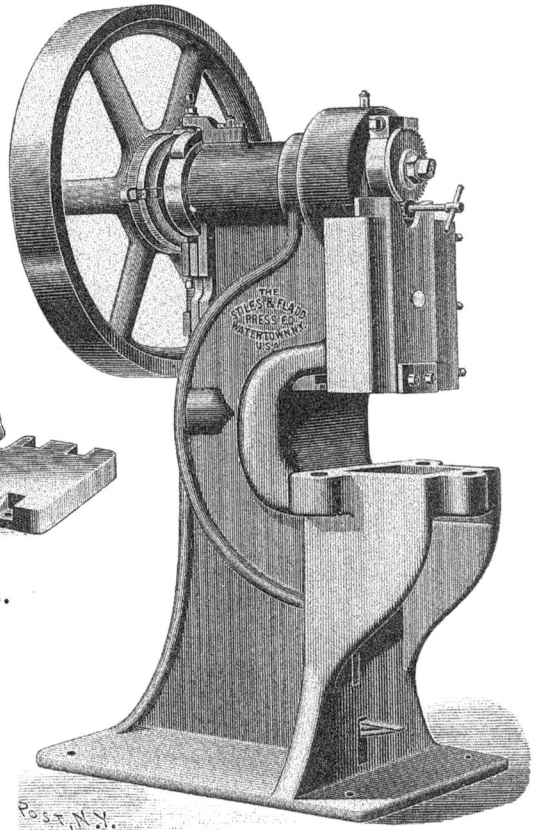

No. 3 Punching Press.

METAL CUTTING, SHAPING and FORMING TOOLS, PRESSES, DIES, DROP HAMMERS and SPECIAL MACHINERY.

WATSON AND STILLMAN, New York, N. Y.

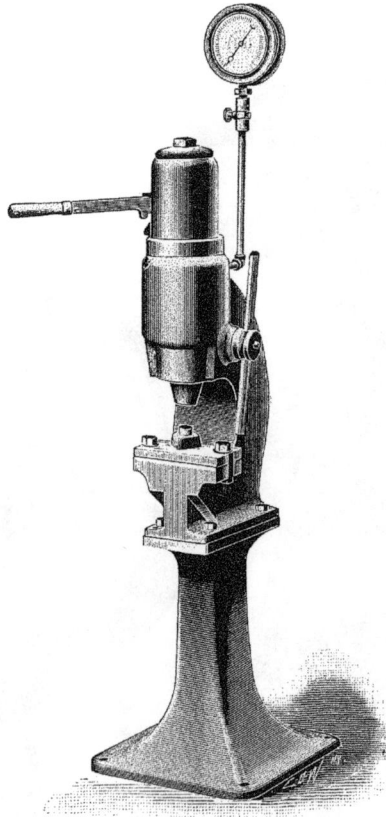

Hydraulic Open Jaw Die Sinking Press.

A very small press with short motion, on which small medals may be struck without the annoying shock of the drop hammer, and is an adaptation of their improved hydraulic punch, strengthened in the jaw to avoid spring.

The body is steel, and from its method of construction irregular dies may be used. The lower lever can, if desired, be made self raising by weighting it at the rear.

Power, 60 to 125 tons pressure.

WATSON AND STILLMAN, New York, N. Y.

Heavy Hydraulic Die Press.

Adapted for hubbing up very heavy dies.

HILLES AND JONES, Wilmington, Delaware.

No. 4 Horizontal Power Press for Very Heavy Work.

For punching boiler heads, crooked furnace plates, bent, angle or
tee iron, boiler legs and all close or crooked work.

G. A. CROSBY & CO., 176 So. Clinton St., Chicago, Ill., U. S. A.

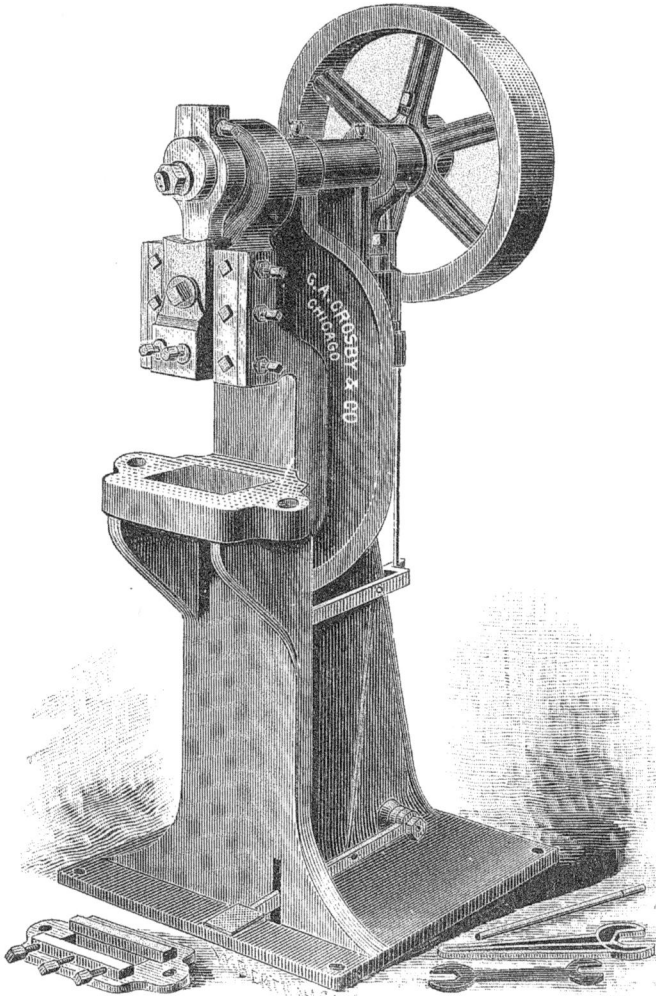

No. 2 Power Punching Press.

Manufacturers of presses, dies, and designers and builders of patent automatic machinery for making cans.   Write for catalogue and information.

THE WATERBURY FARREL FOUNDRY & MACHINE CO.,
Waterbury, Conn.

300 Ton Knuckle=Jointed Embossing Press.

These presses are constructed upon an improved plan which permits a powerful machine to be made in more compact form than formerly. The main frame is made very rigid and as short as possible (by allowing the guides for the crosshead to be placed in front of the knuckle-joints) thereby reducing the spring or elasticity of the frame to a minimum.

A specialty is made of designing and building machinery for specific requirements. Some of our standard machines are as follows: automatic rivet machines, automatic nut blanking machines, thread rolling machines, machinery for making hinges and butts from sheet steel and brass, machinery for making shot-shell and ball cartridges. Patent hydraulic presses, pumps and accumulators.

Power Punching Press P 4.

This press is for heavy cutting, punching, shearing and forming.

12 sizes of this style of press, and a great variety of other designs.

INDEX.

CPSIA information can be obtained
at www.ICGtesting.com
Printed in the USA
BVHW04*1208060818
523683BV00013B/345/P